A Guide to the Electronic Office

The author, Malcolm Peltu, is a consultant on information technology and specialises in interpreting the uses and impact of new electronic and new computer techniques to non-technical people.

He joined the computer industry in 1965 and spent seven years with ICL, Britain's main computer company.

A former editor of Computer Weekly, he is computer consultant to the New Scientist and London editor of the American publication Datamation.

A guide to

THE ELECTRONIC OFFICE

by

MALCOLM PELTU

Based on a study by
The UK Computing Services Association

A HALSTED PRESS BOOK

JOHN WILEY & SONS
New York · Toronto

Published in the USA and Canada by
Halsted Press,
a Division of
John Wiley & Sons, Inc.,
New York

First published 1982

© Copyright M. Peltu and CSA 1981

ISBN 0 470-27308-9

Typeset in 11/12pt Times by
Photo-Graphics, Yarcombe, Nr. Honiton, Devon
Printed in Great Britain by
Biddles Ltd., Guildford & King's Lynn

Contents

[v]

Author's preface

This book is about a technological revolution which is creating a significant evolutionary shift in the nature and efficiency of office work. No technical knowledge is assumed and any necessary jargon is clearly explained before it is used. Although it is of general appeal, the book provides more than just a superficial understanding of the technology. It should therefore be of particular interest to managers and staff who wish to make a practical and reasoned assessment of the way the technology may alter the way information is handled and people work in offices.

Armed with the information and recommendations made in this book, readers will be in a position to face the challenges and risks involved in introducing new technology in their own environment. The advice provided is based on the practical experiences and technological insights of the consultants from the Computing Services Association who undertook a study into the implications of text processing. The CSA study was commissioned by the UK Department of Industry. This book is based on the findings of this study which recommended detailed strategic text processing plans in ten UK organisations. Within the framework of such a strategic plan, the book provides guidelines which should remain valid well into the 1990s, although the capabilities of commercially available technology will change rapidly during that period.

Given this rapid change, it is not possible in a book of this kind to try to offer an up-to-date reference guide to current equipment, prices and suppliers. Within the generalised framework, however, the book provides a great deal of specific and practical advice. For any particular organisation, however, a special and detailed study must be undertaken. The book can act as a first guiding hand but specialist advice should be taken to assist in evaluating an organisation's needs. Such expertise may be found within an organisation or via reputable consultants, such as those who work for organisations who are members of the CSA.

The book is divided into three parts. Part I (Chapters 1 to 3) provides a general introduction to the nature of office information systems. Any reader who already has an understanding of electronic office technology might wish to begin with Part II. The first three chapters, however, provide a coherent analysis of office systems in general and of the technological forces which are being introduced and should therefore be looked at by all readers.

Chapter 1 examines the basic information management functions carried out in every office. Chapter 2 explains how the micro chip has acted as a catalyst in bringing together a variety of electronic technologies. Chapter 3 shows how this new technology is transforming and extending the capabilities of traditional office equipment, such as the typewriter.

Part II (Chapters 4 to 8) provides a detailed look at new office technology. Chapter 4 summarises the main conclusions of the CSA study and Chapter 5 examines the study's findings in more depth. Chapter 6 focuses on word processing. Chapter 7 examines the importance of telecommunications in office systems. Chapter 8 looks at the storing and filing of information.

Part III (Chapters 9 and 10) summarises the main trends and looks to the future. Chapter 9 provides a technological guide to the future in each aspect of an office information system. Chapter 10 discusses the significant long term aim of integrating all electronic office techniques into a single comprehensive system and also examines briefly some wider social implications.

Appendix 1 provides a glossary of terms to help not only with the contents of this book but with the jargon that occurs in newspapers, magazines, sales brochures. Appendix 2 offers a closer look at the nature and findings of the CSA study.

Acknowledgements

The present publication was made possible by the foresight of the Department of Industry in providing the pump primer for the project. The CSA and the members of its Text Processing Group provided the initiative and continuing support to see the project through to a successful conclusion. And Colin Leeson of Langton Information Systems (project leader), Diana Duggan, then of Logica (technical leader) and Willie Jamieson of Arthur Andersen (marketing leader) provided the expert and enthusiastic leadership to manage the project. (See Appendix 2 for more details of the study.)

I would like to thank Colin Leeson for recommending me as a suitable person for turning the study's reports into a book.

For further help there are many word processing, computing and business newspapers and journals which provide up-to-the-minute information on new equipment. Anyone requiring more detailed information about how to apply the technology to a particular organisation could contact:

The Computing Services Association, 73/74 High Holborn, London WC1, Telephone 01-405 3161.

Malcolm Peltu

Introduction

Office automation: the reality

The electronic office revolution will not happen overnight. Changes in handling office information using computer-based techniques will be successful only if they are implemented in a gradual, planned approach.

The CSA study which forms the basis of this book examined what is *actually* happening with office automation rather than indulging in theoretical visions of the so-called 'Paperless Office' or 'Office of the Future'.

Five major themes emerged from evidence found by the CSA consultants:

• *Evolution not revolution*

The reality was found to be less explosive than the frequently painted popular picture of a sudden upheaval that will create a world of automated, people-less offices. This book concentrates on describing the reality.

• *Integrated strategic planning is vital*

In the long term there *will* be radical changes in office information systems. To achieve this, there should be a strategic framework to provide a planned, step-by-step development in which there is a clear understanding of how each task and application relates to and interacts with other office automation activities. This book describes how to develop an imaginative but practical plan to avoid the pitfalls of the office revolution.

• *People before technology*

The effectiveness of new technology ultimately depends on how people understand and use the capabilities, how organisations are restructured to adapt to new techniques and the motivation of *all levels* of staff (office automation affects management as well as clerical staff). This book gives priority to the human factor in office automation.

• *Word processors are the cornerstone*

Microelectronics (the silicon chip) has triggered the development of a wide range of equipment, services and techniques that could be used in the office. The CSA study found, however, that one technique dominated short term office plans — word processing. This book, therefore, describes and emphasises word processing equipment and techniques in detail while also examining other elements in the long term electronic office strategy.

• *Better practical understanding is needed*

The study found that there is still a poor understanding in many organisations of what electronic information systems could achieve in practice. This book attempts to enhance general understanding of the technological potential by placing technical innovation in the context of actual planned developments.

The practical results of the CSA study and the resultant advice and recommendations begin in Part II (Chapter 4) of this book. The starting point is a three-stage action plan:

  • *Immediately* undertake a broadly based study to establish information management, organisational and people requirements.

  • *Plan* long term for at least five years or more with a strategy that shows how various office automation activities fit together into an integrated information management system.

  • *Implement* at once those applications and tasks which fit the plan and can produce immediate benefits.

Part I
Setting the scene

1 Information makes the office world go round

Why work in offices?

Over half of all employed people in industrialised countries work in offices. During the past twenty years, the number of people working in offices has grown while employment in other areas — factories, mines, farms — has fallen.

The type of work carried out in offices covers most aspects of business, social and government life. Whatever the activity, however, all offices have two elements in common: people and information.

People get together in offices to have meetings, answer the telephone, gather and analyse facts about their organisation's work, type letters, file documents, receive and process orders, prepare accounts, use photo copiers, make gossip and generally socialise.

In order to help people carry out these functions in an efficient way, a variety of equipment is used to 'do things' to information. Typewriters, telephones, photo copiers, telex, dictating equipment and computers, all help people to handle information in the office.

Information is gathered, analysed, manipulated, filed, typed, copied, distributed, transmitted, communicated, updated, corrected, generated, lost and found in offices. Information comes in many shapes and forms — as a voice message over the telephone; as written words on a cheque; as text typed on a letter; as words and pictures in books; as numbers on an order form; via normal person-to-person speech and gesticulation.

The ways in which people handle information in offices have changed significantly in the last century, from the time when rows of clerks beavered away with quill pens. Two of the most important office innovations in the last 100 years are the typewriter and telephone, now commonplace in all modern offices. Photo copiers and computers are other examples of what have come to be called 'information technology' because they are technologies which 'do something' with information.

The office way of life

The jobs that people do in offices either fulfil some requirement external to the office, such as being an accountant, marketing manager, social worker or airline booking clerk, or else they are concerned with the internal management and processing of information, such as a typist, data processing specialist, telecommunications manager or filing clerk.

Over the years, people have built up many skills to handle office-based work and have established office routines, organisational structures and job functions to carry out the work. These skills and routines have been developed over many generations, gradually evolving to match the changing social and business environment and to take account of new office information technologies.

This gentle evolution in office work has had to take account of significant but relatively slow technological change when compared to the agricultural and industrial revolutions which were caused by periods of rapid techno-logical innovation. By the end of the 1970s, for example, each office worker was supported by equipment worth only about 10% of the value of the equipment for each worker in manufacturing industry and about 5% of the worth of agricultural equipment per employee.

This book is about how new electronic information technology could create an office revolution, as far-reaching as the radical changes that have altered the nature of agricultural and industrial employment and productivity.

The scope of the impact of electronic office technology is sufficiently important to be called a 'revolution' but the change is unlikely to happen overnight. People, office procedures and organisational structures, and the objectives of people and organisations, usually evolve at a slower pace than the current explosive growth of information technology innovation. Much work still has to be done in translating the technological potential exhibited in experimental or prototype systems into proven, reliable systems that deal effectively and efficiently with the interactions and procedures which ensure that information continues to make the office world go around, even if the office environment of the 1990s may look very different from the traditional office scene.

Before describing the new technologies which will eventually transform the office in the future, it is useful to take a systematic look at the elements that go to make up an office information system.

Functions in an office information system

In many offices, the flow of information may seem a haphazard affair. But even in the most chaotic, ad hoc system there are certain clearly defined functions that are performed: information gathering; storage; processing and analysis; updating; retrieval; communication and distribution; output; management and coordination.

These functions are common to all information systems and have been the basis for the development of particular information technologies and for creating a framework which enables different technologies to be integrated into a new means of information management. Some of the terms used may sound strange but they are just a way of providing a precise and convenient way of discussing activities that are familiar to anyone who has worked in an office. (A summary is provided in Table 1.1 page 12)

Information gathering — collection, preparation and input

Information is gathered and collected in an office by a variety of methods before it is entered (or *input*) into the information system. Information is *collected*, for example, via orders received over the telephone, letters or written instructions, cheques and invoices, personal consultation with a client or speaking into a dictation recorder. Some of this information is entered into the office information system simply by putting a document in a filing cabinet for future reference. Other types of information first need to be *prepared* in some way before being input.

Orders received over the telephone may have to be first written onto order forms before being sent to the relevant department which has to act on the order. Information that is processed by a computer must first be translated into computer code and input onto some device which can be understood by a computer, such as punched cards or magnetic tape.

The act of inputting information to a system may be as simple as filing a document. It could also be entering an item in an accounts ledger or making an airline reservation directly using a device linked to a computer.

Information gathering aims to assemble the facts and figures, background information and historical records which are the information lifeblood of the organisation's activities. Information gathering may be an important activity in its own right or a byproduct of some other task.

Examples of major information gathering tasks are the national census, annual stocktaking, market surveys, and monthly national statistics on unemployment and inflation. Each time pen is put to paper or finger to keyboard, information is being generated, collected, prepared or input.

Information storage

Information needed for carrying out the objectives of an office is stored in filing cabinets, desk drawers, libraries, people's heads, microfilm stores, and computer memory.

Information storage should be designed to simplify access to any item of information or document when it is needed. Urgent information, for example, may be held in the In/Out trays on a desk. Other regularly used information is held in filing cabinets which are organised, say, in alphabetical order of customers or clients, in date order or the most convenient method for that particular information. Information required infrequently or for historical purposes only is likely to be stored in a different way, perhaps on microfilm or in special archives. The sophisticated method of classifying books in libraries according to subject, author, title, is designed to ease access to information.

Information processing

The heart of most office activities is the way information is processed and analysed. Typing a letter or calculating a bill are examples of the processing of information. Pressing keys on a typewriter transforms (processes) information from one form, say, speech or handwriting or thoughts in someone's head, into another form, as printed characters.

Producing a bill illustrates two ingredients in information processing: the performing of numeric calculations and the following of a standard set of procedures that frequently involves reference to other stored information. In the case of a bill, reference needs to be made to information such as prices of products, bulk discounts, special customer discounts, tax rates, customer credit limits.

The production of a bill is primarily concerned with numerical calculations; typing or printing a bill is a relatively simple task. This form of numerically oriented information processing has become known as *data* processing. Information handling that primarily involves words rather than numbers is known as *text* processing, such as letters, memos, reports and general publishing activities. There is a great deal of overlap between these types of information processing, but the distinctions are frequently made to identify the main form of the information involved in a particular task. This book is concerned essentially with text processing.

Text processing can be complex and time consuming. Take, for example, the preparation of a report which must be accurate and well presented in its final typed version. First, the originator or 'author' of the report, say, a manager or research scientist, creates the text in a handwritten or dictated form. This will then be typed as a first draft by a secretary or typist and returned to the author for correction.

Then the author may decide to alter or add to the first draft, as well as correct mistypings. These changes could involve moving sentences and paragraphs around, adding new paragraphs, deleting chunks of text, etc. A completely retyped draft may then be needed, which might introduce new typing errors in previously corrected text. The second draft is then returned to the author and the process continues until the author is satisfied. This could involve the typing of many drafts and the creation of many carbon and photo copies.

Information analysis

Information processing also involves the analysis of stored information. For example, a marketing manager may want to be presented with the latest sales figures and competitors' performance, a production manager may want to know how manufacturing targets are being met or a headmaster may need an analysis of examination results over the last five years. The presentation of such analyses may be in the form of a report consisting mainly of text or as a series of graphs and tables.

Information updating

Stored information must be updated when necessary to reflect accurately the latest situation, unless the information is being stored primarily for archival purposes. Updating may take place as an activity in its own right, say, altering inventory records after an annual stock-taking exercise or altering school records after a major examination. Most updating, however, usually occurs as a byproduct of input and processing activities.

For example, when a customer places an order or makes a payment, the relevant customer record is updated at the time the order or payment is recorded in a book or on a computer system. This information may also be fed through the information system to the production, delivery, customer service, warehousing and other departments to update their own information files and to initiate necessary action.

Different updating procedures and timescales need to be developed to meet varying requirements. The information used when booking an airline ticket or hotel room must be as up to date as possible to avoid confusion and incorrect booking. Sales statistical information, however, may need to be accurate only on a daily, weekly or monthly basis rather than minute by minute. Some historical or archival information can be updated at a more leisurely pace.

Information retrieval

It is often necessary to search through information to retrieve certain documents or to collate particular categories of information. The search is made simpler if the way the information has been stored is designed to match the search routine or if suitable indexes are provided.

The search for a letter from company XYZ dated 5 April last year will be easy to find if the correspondence file is organised in alphabetical order by company and by date within each company. It will be difficult to find if letters are filed in no order at all or if the request is to find something like 'the customer who ordered part number 1234' and no reference to part numbers has been made in the way letters are filed.

A personnel department's information retrieval requirement may be to search through files to produce a list, say, of all female staff over 35 who have been with the company more than ten years. A manager may ask his secretary to book him the lowest-priced flight to New York and then to book him into a reasonable quality hotel close to a particular exhibition centre.

The answer to a question may be 'retrieved' from a filing cabinet, a computer, a person's memory, a book or any other information store. The time for retrieval varies according to the task being carried out. A customer demanding information about the delay in the delivery of an order will want an answer very quickly. The personnel manager trying to find out how many staff live in a particular region in order to help plan some future new office move could probably wait a day or more.

The information retrieval and updating activities must be carefully coordinated because they may have an effect on the same source of information. If information is being updated continuously, the files containing it should not disappear from the filing cabinet for long periods. If certain information is duplicated and stored in a number of files, care must be taken that these files are not updated at different times so that some departments are working with out-of-date information.

Information communication and distribution

Information often has to be sent from one location to another, between people, departments, offices and organisations. Physical copies of information can be distributed using carbon paper, photo copying machines and the internal and external mail system. Communication can, of course, also be made without the physical transfer of a document by, for example, using the telephone or telex. Organising meetings is also a common way of communicating information to many people.

The efficiency of the communications and distribution service plays a crucial role in determining the effectiveness of the overall information system. Just think how office work would be slowed down if there were no telephones to carry certain conversations instantaneously.

The telephone has become the heart of modern office communications. The same network of telephone lines can now be used for other services, such as linking a computer to a device in an office for the direct input and retrieval of computerised information.

Information output

Output can be thought of as what happens at the opposite end of input. The mouthpiece of a telephone, for example, is an input device; the earpiece at the other end provides the output. A typewriter keyboard is used to input the requests of the typist and then the printing mechanism produces the desired output on paper.

Output can take many forms, such as typed and printed letters on paper, pictures (still or moving), speech, messages on display screens linked to computer. In the past, the most common forms of computer output were large stacks of printed reports, called printouts. Now, computers can generate output in a variety of other ways.

Information management

Various activities in an office information system may be the responsibility of different managers. The office manager or administrator could be in charge of basic office services. In addition there may be a telecommunications manager, typing pool supervisor, mail room controller and data processing manager.

As electronic information technology can be used for all these tasks, it is important that there is some form of information management coordination to ensure that all the elements combine together in the optimum way.

Organisation and methods (O&M) studies have traditionally been used to evaluate and design efficient office procedures. In the data processing field, sophisticated systems analysis and design techniques have been developed to create electronic information systems.

The main objective of information management is to use techniques such as these to analyse corporate information needs and to devise the most effective ways of satisfying them. This should be extended to all aspects that affect the cost and efficiency of information handling, even though some activities, such as travel arrangements, may have been regarded as separate issues in the past.

In some larger organisations, an information manager has been appointed to coordinate all information operations. Ultimately, however, the real responsibility for information management lies with senior management who can view the organisation's needs as a whole.

Whatever management structure is adopted, it is important to balance the information requirements of different departments and activities and the overall corporate needs with available information technologies that could range from the familiar typewriter to computer-based devices linked by satellite.

Office information systems: a summary

This chapter has identified the importance of information to the working of any office. It has shown that everyday office procedures and equipment fit into a systematic pattern which can be used to define the information flows and procedures in an office. These office information activities take place using a variety of technologies and equipment.

The basic elements of an office information system are summarised in Table 1.1 page 12. They are presented in a logical order, starting from the collecting of information. In practice, however, the flow is rarely this clearcut, the different phases occur in a variety of patterns and interactions.

Table 1.1 Summary of elements in an office information system

Information activity	Examples
Collection	Taking orders over the phone; receiving letters and cheques through the post
Preparation	Transferring telephone orders onto order forms; turning information into computer-readable form
Input	Filing a document; making an entry into an accounts book; feeding information to a computer
Storage	Filing cabinet; microfilm; computer memory
Processing and analysis	Preparing a bill; calculating staff payroll; analysing census or sales statistics
Updating	Changing inventory records after a stocktaking; putting new documents in a file; deleting a seat on a theatre plan when a ticket is bought
Retrieval	Finding a letter in a filing cabinet; searching a library for particular books or information; looking up a train timetable
Communication and distribution	Making a telephone call; sending a photo copy of a memo through the internal mail; posting letters through the mail; travelling to a meeting
Output	A letter; invoice; report; message on a display screen linked to a computer
Information management	Organisation and methods studies; typing pool supervision; data processing (computing) management

2 Revolutionary forces in the office

The revolutionary technological environment

The techniques of office information systems discussed in Chapter 1 have been adapted over many years to take advantage of innovations in information technology, such as the typewriter and telephone. Until the 1970s, these innovations occurred at a pace which enabled them to be absorbed gradually into office procedures. Each technological advance occurred in isolation from other information technologies. This avoided the creation of a 'revolutionary' environment in which rapid innovation occurred in many activities at the same time, with one advance interacting with another to create a magnified 'shockwave' of change.

In the 1970s, however, developments in electronic information technology began to converge into a unique and powerful force with the potential to create just such a 'revolutionary' technological environment. These electronic innovations encompass every task examined in Chapter 1. In each activity, the changes have taken place rapidly. Because they are based on unifying principles, they have resulted in a total effect far greater than if each development had taken place independently of the others.

The symbol of electronic information technology is the so-called 'micro chip'. Its main significance lies in the way it has become a catalyst for the explosive release of computing power.

At the heart of modern computing and of electronic information technology are two basic concepts:

• *Digital representation of information.* All forms of information (text, sounds, images, etc.) can be translated into a code which uses only the two digits 0 and 1, just as the Morse Code represents the alphabet as a combination of dots and dashes.

• *Program control of information processing.* A computer can perform complex information processing tasks automatically under the control of a detailed program of instructions stored in its memory. By changing the program, the computer can be made to perform many different tasks without having to alter the physical computer equipment. Programs are called *software,* in contrast to the physical equipment, which is called *hardware.*

The rest of this chapter examines these concepts and explains how microelectronics (the 'micro chip') has provided a dramatic impetus to electronic information technology. The aim is to provide an insight into the general nature of the technology and how it applies to office information activities, not to give a detailed introduction to specific techniques.

The use of a telephone does not depend on knowing how the voice signals are transmitted and how a telephone exchange works; such skilled knowledge is needed only by the engineers who develop and maintain the telephone service. The important factors to a telephone user are understanding how to make a call, the ease and speed of making a connection, the reliability of the service, how quickly breakdowns are corrected and, of course, the cost.

In the same way, the user of electronic information technology does not need to know how to write a computer program, design a micro chip or decipher the '0,1' code in order to make use of various devices and services.

However, in order to be in a position to evaluate the true efficiency, reliability, performance and effectiveness of an information system, some insight into the background and nature of information technology is useful.

The roots of electronic information technology

The roots of modern computers can be traced back almost 2,000 years to the Chinese Abacus, which was probably the first computational aid. Through the centuries, mathematicians and philosophers, like the Frenchman Blaise Pascal and German Gottfried Leibnitz, created mechanical machines and other aids to try to automate the performance of basic calculations, such as addition and multiplication. The 'father' of computing, however, is generally recognised as the British inventor Charles Babbage who, in the 1830s, developed a machine known as the Analytical Engine. This introduced the important concept of control by a program which could be input to the machine. Earlier calculating machines had some in-built programmed operations, but were largely manipulated by direct manual operation.

The ability to change the operation of a machine according to the program fed into it meant that Babbage had created a *general purpose* computational machine. He also identified the four other elements which constitute a modern computer: a *processing unit* to perform calculations under program control; *memory* to store information needed during processing and to hold the programs; and *input* and *output* mechanisms to feed information to the computer and to present results.

Charles Babbage used punched cards as the first medium on which to store and input program instructions and other information; Lady Ada Lovelace, his assistant and daughter of poet Lord Byron, prepared such instructions and is regarded as the first *programmer*.

The next major computing innovations occurred shortly before and during the Second World War. In 1936, German Konrad Zuse applied for a patent for an automatic calculator based on the *binary* numerical system, which consists of only the digits 0 and 1 but which is as powerful as our familiar decimal systems with 10 numerals. During the war, the Colossus code breaker in Britain and Eniac (Electronic Numeral Integrator and Calculator) in the US, which was used for calculating trajectories in ballistics research, became the first electronic computing machines. After the war, there were rapid developments in the US and Britain which led to the emergence of the modern computer.

One of the major advances was to store software in the computer's own memory rather than to feed it in each time a program had to be run.

One of the first computers designed for commercial rather than scientific or mathematical tasks was the Lyons Electronic Office (LEO), which was installed in 1953 by caterers J. Lyons and Co. to handle ordering and stock control procedures. This was a turning point in office information systems because it showed that what had been conceived previously as a device whose natural habitat was a research laboratory could have a valuable role to play in bread-and-butter commercial tasks.

These roots of computing can, however, be regarded as a long, slow burning fuse leading to the micro chip trigger. Another long thread interwoven in the roots of electronic information technology concerns telecommunications, the transmission of information.

Getting the message across

While computing developments in the 19th century drifted in a mathematical backwater, major practical advances were being made in telecommunications. In 1837, Englishmen Cooke and Wheatstone invented the telegraph and the first telegraph line was laid in 1843, between Paddington and Slough for use by the Great Western Railway. Then, in 1876, a Scotsman, Alexander Graham Bell invented the telephone and at the turn of the century, an Italian, Marconi introduced wireless transmission.

A great boost to the use of the telegraph was the invention by American Samuel Morse of his Morse Code. This is based on dots and dashes, a dot being transmitted as a short electrical signal and a dash by a longer signal. The alphabet and digits 1 to 10 were represented by a sequence of up to five dots and dashes.

Later, a more comprehensive 0,1 binary telegraph code was created by Frenchman J.M.E. Baudot which utilised a punched paper tape to send and receive messages — the 'ticker tape'. The tape has five channels in which a hole is punched to represent '1' or left blank to represent '0'. Each row on the tape thus carries the code in five binary digits. The letter 'D', for example, is encoded as 10010. The *teleprinter* was developed which accepted signals sent in this code and printed them automatically. Telex service was started in the 1930s which used typewriter keyboards for input, punched paper tape for message storage and the teleprinter for output.

The transmission of voices and text messages by telephone and telex is only half the technological picture. The other main telecommunications technique concerns the switching of calls so that many people can use a network and dial up any other receiver. The first telephone exchange was a manual switchboard, the operator connecting two callers by plugging the two ends of a cord into the appropriate sockets on the switchboard.

In 1889, an American undertaker, Almon B. Strowger, developed the first automatic exchange because he believed that corrupt telephone operators were transferring his business calls to rivals. He used electro-mechanical techniques to select the appropriate line based on the pulses generated when a number is dialled. Strowger exchanges with clunking mechanical switching were still used by over 85% of the UK telephone network in the mid-1970s.

Wireless, TV and satellite communications extended the ability to transmit a variety of information over long distances.

The 1960s also saw a commercial boost being given to a technology which was originated in 1843 by a Scottish inventor, Thomas Bain. His device aimed to transmit a copy of a document or photograph from one location to another. During the first half of the 20th century, this capability was provided in 'wirephoto' services by news agencies and in the transmission of weather maps. In the 1960s, a number of manufacturers brought out products which were known as *facsimile* transmission devices (or *fax* for short). A document can be placed in a fax transmitter and a copy produced on a receiver at the other end of a telecommunications link.

Computing and telecommunications come closer

Although the development of computing and telecommunications began in different directions, they have now converged onto the same path, united by a common adherence to the two basic electronic information technology principles:

- digital information representation
- automatic program control of information processing

Traditionally, telecommunications transmissions were performed using *analogue* techniques. Telephone voices, for example, have been sent as analogue wave patterns that travel along the telephone line. However, digital transmission has been found to be more reliable, efficient, flexible and cost effective than analogue methods.

At the same time, the switching operation has been taken over by computer-controlled exchanges, such as British Telecom's System X, which offers enormous price and performance advantages over previous electro-mechanical exchanges.

Computer communications systems have always been based on digital techniques, but have had to take account of the analogue telephone network by using special devices, called *modems*, to translate between the digital and analogue forms.

Computer data and telephone voices are not the only types of information that can be handled by a digital communications network. In electronic information terms, the differences between, say, a page of A4 text, a telephone message or a TV picture are the numbers of binary digits (called *bits*) required to convey all the information. The more bits that are needed, the greater is the bits per second (bps) rate required to transmit the information.

A telex message, for example, can be sent at a speed of about 50 bps; medium-speed computer data transmission travels at about 4,800 bps; telephone speech at up to 64,000 bps; and colour TV at over 90 million bps.

Information travelling along a telecommunications link can be thought of as water flowing along a pipe. Just as a wider pipe allows a greater volume of water to pass through it more quickly, so the *bandwidth* of telecommunications channels determines the volume of information flow.

Analogue transmission makes inefficient use of bandwidth because the transmission wave forms overlap and cause interference. Digital channels can be packed more closely together and can make optimum use of new *wideband* (sometimes called *broadband*) transmission media, such as optical fibres, microwave and satellite communications.

A traditional telephone copper cable can carry about 1.5 million bps. This could be divided into, say, 24 telephone conversations each of 64,000 bps or a greater number of telex channels but would be too narrow to carry TV pictures. Optical fibre links, which transmit information via laser-generated beams of light, can transmit over 1,000 million bps and a single satellite communications channel can handle tens of millions bps.

This means that the same communications network can handle a mix of all forms of digitised information by using broadband links with computer-controlled switching. By combining this with computer-based techniques for storing information in computerised libraries and processing information, a powerful new technological force has emerged which the French call *télématique*.

In English, the terms 'information technology' and 'computer communications' are used to describe these developments. In any language, however, the convergence of computing and telecommunications has produced a range of exciting new technological opportunities.

The incredible shrinking machine

The earliest computer systems filled whole rooms. This was because they used thousands of valves (just like those used in radio sets) for the processor and a variety of cumbersome mechanisms for storage, such as cathode ray tubes and five-foot tubes of mercury. Then the micro triggered the story of the incredible shrinking computer.

Processor progress

Each processor operation, such as addition or multiplication, consists of many smaller steps which are based on a system of logic devised by the 19th Century British mathematician, George Boole. Like the binary system and Morse Code, Boolean logic is based on a 'two state' representation; in this case 'true' and 'false' are the two possible values.

A computer processor consists of circuits made up by inter-connecting 'switches' in logical patterns defined according to

Boolean logic rules. Originally, valves acted as the switches in a circuit; the value of a switch is regarded as 'true' or 'false', depending on whether the valve is on or off.

Valves were later replaced by transistors, just as transistors replaced valves in radio sets. (In radios, valves and transistors are used to amplify sounds; in processors they act purely as switches.) Although transistors perform exactly the same functions as valves, they are much smaller, more reliable, faster to operate, very much cheaper to produce and consume much less energy.

The transistor was developed at Bell Laboratories in the US shortly after the Second World War, but only became available in 1956. Transistors are made of *semiconductor* materials, such as a constituent of sand called *silicon*.

As its name suggests, a semiconductor is halfway between being a good conductor of electricity and an insulator. Computer processor circuits can be created by introducing impurities into the semiconductor material according to the required circuit patterns.

At first, only one transistor could be made on each sliver of semiconductor material. The transistors were then placed together with other components onto boards and connected by copper wires to create processor circuits. *Integrated circuits,* where more than one transistor is integrated onto a single silicon *chip,* soon followed. The ability to pack transistors onto a chip which is only a few millimeters square, grew rapidly. In 1959 there was just one transistor per chip. Since then, the number more or less doubled every year. By 1980, tens of thousands of transistors could be placed on a chip. The degree of compactness has been identified by terms such as medium scale integration (MSI), and large scale and very large scale integration (LSI and VLSI).

The first *microprocessor* came in 1971 when the Intel Corporation produced the first complete computer processor on a chip containing over 2,000 transistors.

Memory march

Transistors also transformed another key aspect of a computer, the memory or storage. *Main memory* is used to store programs and information needed immediately to perform a task. Other storage is used for larger volumes of information that are required less urgently and can be read into main memory when required; this type of memory is called *mass, bulk* or *backing* storage.

Media that can distinguish between two states can be used for computer storage. In punched cards and paper tape, the two states are the absence or presence of a hole. Information in magnetic tape and discs can be represented by the direction of tiny magnets in the magnetic coating. Transistors can act as on/off two-state switches.

Microelectronics has made its major impact on main memory, where speed of access to information is vital, given that it must keep pace with processors that perform operations in a fraction of a second. Alternatives to magnetic media for mass storage are becoming available, but magnetic media are still the most popular (See Chapter 8).

In early computers, valves and cathode ray tubes were amongst the primitive forms of main storage techniques. In the 1960s they were replaced by *core stores* which consisted of many magnetic cores that could be magnetised in one of two directions. Core stores were so popular that the term 'core' is still sometimes mistakenly used to refer to main memory.

Just as transistors replaced valves in processor circuits, so did they replace core stores in memories and provided similar benefits: high speed, low cost, compactness and reliability. The march of silicon memories has been signposted by the number of bits that could be included on a chip, starting with 1024-bit chips (1024 is often abbreviated to K; the number 1024 has significance in binary arithmetic).

By the beginning of the 1980s, 4K and 16K-bit memories were commonplace, 64K chips were appearing and 256K memories on a chip had become feasible. Semiconductor chips have virtually replaced all other forms of main memory.

A volatile cocktail

Microelectronics has therefore shrunk the size, lowered the cost, increased the reliability and improved the performance of computer processors and memories. This has taken computing power to places where computers had never gone before — into watches and washing machines, typewriters and photo copiers.

It is possible to put all the major components of a computer — processor, main memory and input/output interfaces — onto a single chip.

When added to the advances in computers and communications already discussed, the silicon chip creates a volatile information technology cocktail.

The technological forces march in unison

This chapter has examined the three main ingredients of electronic information technology:

- *computing* which enables information to be stored and processed in digital form under the automatic control of stored program software

- *telecommunications* which has moved towards digital transmission and computer-controlled switching

- *microelectronics* which has cut the cost and size of computers and greatly increased their potential areas of cost-effective application

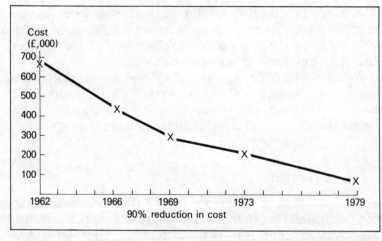

Figure 2.1   How total costs for comparable largish commercial computing systems fell between 1962 and 1979

(*Source: Computer Weekly*)

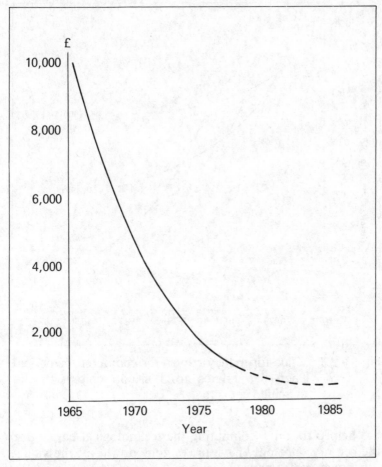

Figure 2.2 The falling curve of prices for the basic equipment and operating programs of smaller computers, called minicomputers

Although these technologies emerged from different roots and were originally developed independently of each other, they have now grown together into a single, coherent integrated technology which can tackle all the office information tasks discussed in Chapter 1.

The importance of microelectronics in bringing computer power into widespread use is summarised in Figures 2.1 to 2.3 which provide just three indications of how the silicon chip

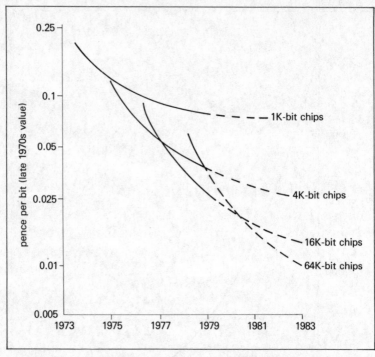

Figure 2.3   Value-for-money curves for computer memories. Each line relates to a silicon chip with the specified capacity

has helped to make computing the technological bargain of the century. Not only have prices gone down but the value-for-money performance has gone up.

Some of the elements of the electronic office of the future have already taken up position, like advance guards of a revolution. Computers and word processors, which have been used in some offices for many years, are just two examples.

Many apparent old codgers of the office, like the telephone and telex, will also suddenly take on vigorous new life when infused with the digital spirit of electronic information technology.

3 From typewriter to electronic city

The electronic office takes shape

Information technology is not new. From the time of the first cave paintings people have been using techniques to communicate information. The digital electronic information technology described in the previous chapter is relatively new, but did not suddenly appear out of the blue with the invention of silicon chip integrated circuits.

Most of the equipment and techniques which are likely to be commonplace in what has been called the *Electronic Office* or *Office of the Future* are developments of technologies and equipment that have been known about for most of the 20th century.

The best way of looking at the electronic office is as a jigsaw puzzle. Each piece in the puzzle performs a particular information task. The technology used for one piece could be some traditional technique, like a typewriter keyboard, or it may involve some new electronic technique, like having a screen instead of paper as output.

Figure 3.1 page 27 shows the link-up of components in this office jigsaw. Each row in the figure represents one information management task. There are, of course, many other elements that could be part of the overall picture, but the most important information activities have been identified: input, storage, processing, communications and output.

This chapter examines how the main elements of electronic office technology have developed from traditional equipment and techniques.

The typewriter to word processor evolution

Originally the typewriter consisted of two basic functions: the keyboard (for input) and printing mechanism (for output). Until the late 1960s, the main innovations were directed at speeding up the whole typing process and to make the output more flexible. The electric typewriter, for example, cut the time between touching a key and the character being printed on the paper.

The use of a single print element 'golfball' head replaced mechanical print arms, which not only made typing quicker but gave flexibility in enabling the typeface to be altered by changing golfballs. More efficient single print elements have been developed to improve printed output techniques, such as the *daisy wheel* or *petal disc*, where the type characters are placed around a moveable wheel. Although these single element output mechanisms became commercially popular only in the 1960s, the first single element electric typewriter, the Blickensderfer, first appeared in 1902.

An important innovation in typewriter technology was the introduction of storage (memory) techniques such as punched paper tape and, more recently, magnetic cards, tape and floppy disks (which resemble flexible 45 rpm music records.) Input typed on the keyboard is stored in digital form on this storage, in addition to or instead of being output onto paper.

The first automatic typewriters with memories appeared over 50 years ago, gained popularity in the 1950s and became widespread in the 1960s with the advent of the IBM Selectric magnetic tape machine. With such a machine, the typist can perform an information processing task called *editing*. This enables corrections to be made to typed drafts using automatic aids. For example, when a draft of a letter has been stored on magnetic tape and corrections need to be made, the typist 'replays' the tape, making corrections at the appropriate points and storing the corrected letter in the memory.

After the automatic typewriter with memory, the next information management function to be incorporated in or linked to a typewriter was a computer processor, which led IBM to coin the term *word processor*. At first the software programs and the amount of memory available allowed only

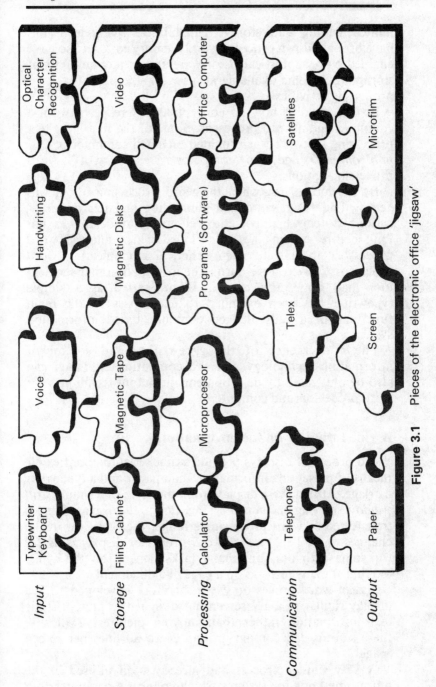

Figure 3.1   Pieces of the electronic office 'jigsaw'

limited editing and storage capabilities. Microelectronics have brought down processing and memory costs and so have enabled more sophisticated software and greater amounts of storage to become available at a reasonable price for word processing activities.

In fact, there is now no conceptual difference between a computer and a word processor. Some of the first large text processing software was provided on small computers called *minicomputers* and many computer systems now offer word processing options.

By the late 1970s, when the word processing boom was getting underway, combined computing and telecommunications systems had become well understood. It was not long before systems were developed which allowed word processors to talk to each other. Communicating word processors (CWP) were born. The information management loop had now been closed. What started as a simple typewriter has been extended to encompass all the main functions of a fully fledged computer-based information system.

The convergence of the typewriter-based stream of information technology and the computing approach can also be illustrated by developments in output techniques for word processors and computers.

A video window on the information world

Word processor displays provide a quicker and more flexible means of presenting information than paper and a typewriter carriage. At first these displays took the form of a simple *strip window* showing one line of typing. Before the displayed line is transferred to word processing storage, the typist can make changes to it. This type of word processor display is useful for some activities, but many tasks depend on the typist looking at many lines, as on a page of a document.

Screen word processors were therefore developed with a display similar to a TV screen but designed to present static text information rather than moving pictures. Typically, these screens can contain around 20 to 30 lines per screen 'page'.

A very similar process had already been utilised in the output methods for systems which connect a computer to a device at a distant location. The earliest form of such direct online computer communications used a teletypewriter for the main output device. This is similar to the teleprinter used with telex services and popularised by the teleprinter results service which used to chug out football scores at 4.40 p.m. on Saturday TV sports programmes.

The teletypewriter proved much too slow to cope with the high speed at which computer data can be transmitted. It was therefore replaced by a TV-like screen which was much faster and more efficient than teleprinters. On Saturday sports TV, for example, a 'videoprinter' is now used which is quicker and quieter than the old teleprinter.

In addition to receiving output from a computer over communication lines, online systems were also developed to enable information to be input directly. The teleprinter had a typewriter-like keyboard attached to the printer. Its screen-based replacement similarly has an associated keyboard. The whole screen-plus-keyboard unit is generally called a VDU (visual display unit) in the UK although the term VDT (visual display terminal) is often used elsewhere. A *terminal* in this context means any device linked directly to a computer.

Returning now to the screen-based word processor, it can be seen that it is a form of VDU because it has keyboard input and screen output. Word processors can be linked to each other to transmit information, just as VDUs can be interlinked to exchange information and messages. Both VDUs and word processors can incorporate their own processing and storage capability or a number of devices can share processing and storage through a computing unit in the office or via a telecommunications link to a computer.

The way in which word processors and VDUs emerged independently from different technological 'cultures' is a story that is repeated time and again with electronic information technology. The word processor came from the culture of typewriter-oriented office equipment whereas the VDU came from the computing data processing world. As they are based on the same principles of programmable digital information handling, they naturally converged until the distinctions become meaningless except in terms of what tasks they perform.

Every image tells a story

Typing and word processing involve handling text. A great deal of information, however, is in the form of some kind of graphical information — pictures, sketches, architectural blueprints, graphs, bar charts.

The technique of xerography was developed to enable copies of documents to be made without going through photographic processes involving the creation of negatives. This produced the well-known photo copier which has become a common feature of most offices.

The photo copier is an example of what is called *image* processing. It fulfills the input-processing-output part of the information management cycle: the original document is fed in as input, an image copying process is then performed on it and the final copy is output.

Another form of image processing is facsimile transmission. This satisfies the communications as well as the input, processing and output functions of information management.

There are also devices called *digitisers* which are used to trace over a blueprint diagram so that the design is transformed into digital form for storage and processing on a computer. It is now possible to 'draw' directly onto a screen with a special *light pen* with the resultant image directly translated into digital form.

Image processing can therefore be performed as a purpose-built process, such as in a photocopier, or the image can be translated into digital code and be integrated into electronic information systems.

Letters down a telephone line

There was a time when tom-toms and smoke signals were the quickest means of sending a message from one point to another. As travel became easier, mail services were developed to transport information and objects physically.

Some mail consists of physical objects. However, a great deal of it is essentially the communication of information. This communication could be done by electronic means rather than through the post, if suitable *electronic mail* facilities are available.

'Electronic mail' is a term that was invented relatively recently to describe capabilities as old as the telephone and telex. Electronic mail means the transmission of information between one place and another by electronic means instead of by physical mail methods. A telephone call, for example, may be made instead of writing a letter. Telex and facsimile could also be regarded as electronic mail.

Electronic information systems are providing more comprehensive and powerful electronic mail capabilities. Communicating word processors can exchange letters, memos and reports, untouched by human hand. Goods can be ordered directly from a computer terminal instead of writing a letter or posting an order form. Payments can be made directly by electronic transfers of funds between computerised bank accounts instead of posting a cheque.

A home improvement idea for the office

So far, this chapter has illustrated how traditional office equipment has blossomed into electronic information technology devices and services. Developments in other areas have also been drawn into the office information net.

The Prestel viewdata service, for example, was first introduced by British Telecom in the late 1970s as a means of giving access to computerised libraries from a domestic TV set. With the addition of a viewdata adaptor and a small numeric keypad, the TV set becomes equivalent to a VDU, with the keypad used for input and the screen for output.

Prestel was designed to be easy to use and to provide an efficient means of setting up a computerised information base. The viewdata capabilities pioneered by Prestel quickly proved to be attractive to organisations for internal and external communications and for connecting to external information services, such as Prestel itself. Private business viewdata systems were developed, including viewdata terminals that resemble VDUs more closely.

Viewdata has, therefore, become another part of the electronic office, although it started out as a service for the home.

Glueing the electronic office pieces together

The various information technologies that contribute to electronic office systems have been derived from a variety of sources and are supplied by a wide range of suppliers. Theoretically, it is feasible to integrate these systems into a single service.

The telephone network is an example of how communications problems can be overcome through firm standardised policies. International compatibility procedures must be implemented to enable a person to pick up a telephone and dial a number in another country. There must be agreement, for example, on dialling codes and the way voice signals are sent and received.

This kind of service is taken for granted with telephones. It is more difficult to organise with the variety of electronic office systems and equipment emerging onto the market.

Various makes of word processor, for example, have different internal codes to store and transmit information which are incompatible with each other and with computer systems to which a user might wish to link them.

In order to overcome these communications compatibility problems, techniques have been developed to handle *networks* of devices. These networking techniques involve overall architectural plans which act as the blueprint for interconnection paths and which define the standards (*protocols*) that must be adhered to by devices which 'talk' to each other.

Having individual office information pieces is insufficient; there must be an adequate communications network to glue them together.

All aboard for the office workstation

This chapter has been about convergence and divergence. The convergence of various information management techniques into a unique, integrated blend of electronic information technology. From the point of view of particular devices and techniques, however, there has been a divergence from, say, the relatively simple function of an ordinary typewriter to the complex and powerful capabilities of a word processor.

A symbol of these developments is the coining of the term *workstation* for the unit at which office work takes place. In its most basic form, a workstation is a desk plus typewriter, telephone and writing equipment; it could also be a word processor or a VDU. A workstation can be thought of as a point from which a variety of departures can be made to carry out different types of activities, and to send and receive information.

At the end of the line, so to speak, is an integrated, multi-function workstation which could eventually incorporate a desk, keyboard, screen, fax, telephone, direct voice and handwriting input to a computer, some form of compact mass storage, access to an electronic mail box service, public and private viewdata systems and other capabilities.

Such an all-singing, all-dancing workstation is unlikely to be an overnight technological sensation. The forces motivating the office 'revolution' are moving at a slower, more human pace.

How technology meets office information needs

This section of the book (Chapters 1 to 3) has placed developments in office information systems into historical perspective, both in technological terms and in the evolution of traditional office procedures.

Table 1.1 summarised the major elements in any office information system. The table on the following page is a brief summary of devices and systems that can meet these requirements.

The rest of this book explores the realities of the future of electronic office systems.

Table 3.1 Summary of the electronic information technology devices and systems in an office information system

Information activity	Examples
Information collection	Orders via VDU; electronic mail
Preparation	Computer data preparation; filling in forms for direct input
Input	Word processor or VDU keyboard; direct; direct handwriting or voice input to computer-based systems
Storage	Silicon chip main memories; magnetic disks and tapes; video disks
Processing and analysis	Software; word processor editing
Updating	Software; online VDUs
Retrieval	Software; computerised libraries; viewdata
Communication and distribution	Computer networks; communicating word processors; electronic mail; facsimile; intelligent copier
Output	Screen, printer, computer-controlled voices
Information management	Network architectures; communications standards; corporate information studies

Part II
Today's office
landscape

4 How people can keep control over technological power

People are more important than technology

The silicon chip has become a symbol of the electronic information technology changes discussed earlier in this book. It is an apt technical image for the many convergent technologies that have been given such a decisive push by the micro, yet it is an inappropriate representation of the wider human, organisational and systems factors which are vital in determining how the technological potential is harnessed to tackle specific tasks.

The Computing Services Association (CSA) Text Processing study examined the realities which will decide where, how fast and how far technical theory will be put into practice. The results of this study, which form the basis of the rest of this book, draw a clear picture of the likely organisational impact of the technology:

- People are more important than technology
- Some technical and organisational change is inevitable
- This change will be at an evolutionary not a revolutionary pace
- The rate of change in offices will be limited more by the ability of people to assimilate change than the availability of raw technology
- Staff at all levels of an organisation will be affected
- Organisations must plan for the long term but implement at once any application of the technology that can produce real benefit immediately.

THE FOLLOWING IS A SUMMARY OF THE MAIN FINDINGS OF THE CSA STUDY:

Managing the human factors

The future offers a certainty of new opportunities and new risks related to the way we produce and use information. These opportunities are of a size which may eventually affect the viability of a company as well as the prospects of its staff.

There is no sure way of avoiding problems in introducing new office technology. A reluctance to accept the technology may mean losing the ability to compete successfully in the marketplace for sales or staff. On the other hand, over enthusiasm for the new technology without adequate understanding can mean costly errors, staff dissatisfaction and early equipment obsolescence. It is therefore important that decision makers in an organisation have a broad understanding of how technology might change their needs for equipment and staff, their organisational structures and information management procedures, job design and a variety of other human behavioural and organisational effects.

The new systems must be designed to help people rather than trying to change their behaviour just to suit the system. Automating an inefficient operation does not necessarily make it efficient.

People and organisations take time to understand and assimilate change, particularly when there are many technical innovations which influence a wide range of activities at the same time.

Many people who are aware of electronic office technical developments may feel that they will have an impact only on typists, secretaries and clerical staff. This is because the word processor was the first swallow heralding the technological summer. In the long run, however, senior staff at all levels will also be affected. Work on systems designed to improve management productivity has shown the feasibility of such an extension of the scope of electronic systems and there are some clear economic benefits to encourage further developments.

Typically, typists and secretaries represent less than 10% of office staff costs. Their influence on company profitability is likely to be even smaller because they are not the decision makers. Applying new technology to improve the efficiency of professional and managerial staff may be even more beneficial than applying it to typists and secretaries because the senior staff have a greater impact on organisation cost, effectiveness and, where applicable, profits.

Technological innovation will lead to organisational changes, for example, in the way responsibilities are grouped and the division of work between locations. There will also be new environmental, job design and ergonomic requirements to maintain adequate levels of job satisfaction, staff health and efficient and reliable equipment operation.

Working relationships between managers and office support staff (secretaries, typists, clerks) will certainly have to change in order to ensure that management effectiveness is improved, as well as increasing the cost-effectiveness of the support function. This is likely to lead to a substantial move in offices away from personal secretaries towards small integrated groups of support staff providing a full range of administrative services for groups of managers.

These administrative support centres will take over more of the managers' routine work while more of their own office tasks are in turn taken over or eliminated by new equipment. Such work reorganisation must also be accompanied by changes in attitude if they are to be effective. Management should create opportunities to encourage a positive staff response to these changes. They should aim wherever possible to provide opportunities for greater job satisfaction and increased pay through improved productivity and to provide additional employment where increased competitiveness leads to expansion.

Some existing jobs will inevitably disappear or alter and the growth in the number of office workers may be halted. Managers are among those likely to be faced with great change, including the apparent drop in status from losing a personal secretary and the increased use of workstations requiring the need to learn 'keyboard' skills.

Given the probable evolutionary pace of developments, it is unlikely that using new technology in offices will result in sudden large scale reductions in the numbers of staff employed. Although typing productivity of 250% and more can be achieved using word processors, such savings are not necessarily converted directly into a reduction of staff because some of the time saved will be used up in improving the quality of typewritten output and partly dissipated by taking on additional work which was previously impossible or unsuitable for support staff to do.

The introduction of text processing equipment into an organisation can be a major project requiring special management skills and coordination between managers and staff in many fields. The implications for organisational and personal change mean that technological innovation must be regarded as a systematic task requiring senior management attention and involvement.

Technological forces for change

There will be major technological changes during the 1980s in the way we handle and provide information. The underlying cause of the change is the continuing, at times revolutionary, improvements in electronic systems such as those using microprocessors and lasers. Prices are tumbling while capabilities are rocketing.

Despite the revolutionary nature of some technical changes, the overall change will be evolutionary and take place over many years. Although many separate pieces of equipment exist which could form the basis of an integrated electronic office, most of them work to very different standards and a vast amount of work is needed to make them fit together.

In the UK in 1980, about 90% of the office population had never used any of the new text processing equipment. Until there is wider experience over a reasonable period of time, it will be difficult for people to express accurately what they need from a new system. Therefore it is very difficult to produce a good design for an integrated electronic office information system in the detail needed to produce comprehensive software.

The best scope for advanced uses of new technology discovered in the CSA study were in the organisations that were already among the most experienced users of text processing equipment. The experience gained was a springboard which helped their staff to appreciate new opportunities and quantify and justify the implications of more advanced projects. This is why an evolutionary approach is likely, although the pace may quicken when more organisations gain experience.

The cornerstone of most developments in text processing will be the word processor. Some convergence between text processing, data processing and communications activities must occur though the path will not necessarily be smooth. There is a great deal of similarity of concepts and equipment used but often the driving forces and attitudes are very different.

Data processing departments, for example, have often been based on large computers which encouraged the centralisation of its activities and have been staffed mainly by technical specialists. Text processing, however, is developing in office environments where the need for centralisation does not apply, because micros have cut equipment costs and management and staff have little technical computing know-how.

Many computer systems adopted a centralised strategy primarily because of the cost and size of early commercial products. There is no longer a significant technical or cost constraint. Organisations must plan their approach according to the real needs of the organisation and of each department and activity. The importance to the organisation of safe-guarding its vulnerability to sabotage, natural disasters, industrial action and major equipment failures must also be taken into account; a completely centralised system could be vulnerable to such events but so could a more decentralised system if units remote from the centre cannot function independently when there is a central failure.

Technically, the two most significant forces pushing for change are the *falling costs* and *increasing capabilities* of electronic information technology. Although these forces are strong, care must be taken when evaluating the real cost/benefit equations.

The most dramatic reductions have been in the prices of electronic components, particularly computer memories and microprocessors. Total systems costs, however, have fallen less dramatically. Some price reduction indicators are given in Chapter 2. Although there are great variations, particularly in evaluating costs of complete systems, by 1980 prices of components dropped roughly to *one-hundredth* of their value in 1970, whereas systems prices during the same period fell to about *one-tenth* of their 1970 prices. If other costs are taken into account, such as staff training, management time, new technical staff, the impact of organisational change, etc., the margin of difference between system and component cost changes would be even greater.

There are two reasons why system costs have reduced less than those of components. Firstly, there is a tendency to use hardware components more extravagantly as they get cheaper, particularly if this reduces other costs. Secondly, complete systems are produced in smaller numbers than components and often carry a high staff overhead cost, which includes software, marketing and other developmental activities. As there is generally a shortage of staff with appropriate technical experience, the people costs in systems have been rising particularly steeply.

In some systems, the component costs are only about 10% of the total, which means there is still scope for reducing systems costs even if price performance trends for components were to tail off.

The main new capabilities of electronic systems relate to their compactness, robustness, speed and software.

The compactness and robustness brought about by microelectronics makes it possible to distribute computer-based equipment more widely and to take computing power to where the work is rather than vice versa, as happened with large and expensive centralised computer centres.

The speed of electronic information processing and communications brings different locations closer together to communicate messages, documents, give answers to queries, almost instantaneously, if necessary.

Software is probably the most significant development because it is this which gives microelectronic versatility and adaptability to a variety of tasks that people wish to carry out. For example, in order to give computers the ability to recognise human speech and handwriting, complex software programs must be devised to control the processing.

A plan for action

In order to ensure that the people and technology interact successfully to the benefit of the whole organisation, a three-stage action strategy should be developed:

- *Undertake a broadly based study immediately* to establish the information handling needs of the whole organisation. Assess how current and expected electronic information technology equipment might satisfy these needs and whether it can be used to create new business opportunities.

- *Produce a long-term information management plan covering five years or more* showing how various activities will fit together. Establish detailed strategic policies and standards where appropriate, including organisational requirements, the impact on all levels of staff, guidelines on the purchase of new equipment and a planned, step-by-step approach to joining up the basic 'building blocks' of equipment and services. This plan will almost certainly need amending as time passes and should be regarded as a flexible framework rather than a rigid specification.

- *Implement at once* those applications and tasks which fit within the overall plan and can produce real benefits immediately through the introduction of some form of new electronic information technology. Start with fairly small projects, take advice from people with previous practical experience and only implement one project at a time.

The equipment chosen for most initial electronic office projects is likely to be based on a relatively simple word processor. The aim of the early projects should be to give management and staff experience in using and assessing equipment as well as providing any direct benefits and costs savings. Although prices have been falling, it is likely that the amount of equipment that is installed will be limited by its cost, which may be greater than any potential benefits.

In the CSA studies, for example, many organisations were recommended to get a number of word processors which meant that much of the typing work would continue to be done on electric typewriters in the short and medium term.

To get more word processors could not be justified.

As it is important that a system can grow and be extended from relatively small beginnings, organisations should judge suppliers in terms of their overall business stability and commitment to electronic office developments, as well as any characteristics of a particular product. Many organisations will prefer to deal with a single supplier for a wide range of equipment and services.

The objective of the planning exercise is to optimise overall information management to benefit the whole organisation. The aim is *not* to introduce glittering new technology for its own sake, just because it seems a modern thing to do.

5 Office information systems—the reality

Putting flesh on the technical bones

The starting point for developing and implementing plans for introducing new office technology must be based on a good understanding of the real needs of an organisation. The CSA Text Processing study was commissioned by the Department of Industry in order to flesh out the theoretical bones of technological potential with practical experiences and needs. In addition to carrying out background research into technical and product developments, the CSA consultants carried out a strategic study which examined the text processing needs of ten organisations in the UK. They made detailed proposals for each organisation, covering the next five years at least.

An analysis was made in each organisation of the current text processing activities and the overall strategy was divided into short, medium, and long term plans.

The study team took a broad definition of text processing as encompassing all ways of using technology to handle information which is currently, or could be, expressed as text. This extended the inquiry beyond just word processing needs and overlaps with wider concepts of the electronic Office of the Future and traditional computer data processing. In some respects, the study's definition of text processing goes beyond the office because it includes activities which do not necessarily take place in an office, like a salesman recording an order. On the other hand, its main emphasis does not cover all office activities as it was focused on systems where some form of 'intelligent', computer-based processing takes place rather than, say, just simple reproduction of information.

Instead of trying to gather a relatively superficial level of information from more companies, the study decided to tackle the ten organisations in depth as a representative national sample in terms of size and area of office activity in the private and public sectors.

No generalised study can replace a detailed examination of each organisation's particular requirements. The conclusions of the CSA study and the advice in this book do, however, provide useful guidelines for developing specific strategies. The general impression gained by the study matches the experience of many users and experts in this field and so can be taken for wider applicability than the limited sample might suggest. Further details of the results are given in Appendix 2.

A snapshot of office needs and equipment recommendations

Current information practices and workloads and future objectives were investigated in each organisation and specific equipment was recommended over three phases from a short-term of about 18 months to longer term strategies of over three years ahead.

Word processors: the basic building block

In all ten organisations, consultants could see areas where more electronic text processing could be usefully undertaken even when there were very few typing and secretarial staff. Word processing was recommended as the cornerstone of wider electronic office strategies in nine organisations. From this first word processing 'building block', the long-term strategies usually involved integrating various information handling systems. In some cases, *executive workstations* were recommended to improve the productivity of senior professional and managerial staff.

All typing and secretarial staff were found to spend a little over half their time typing; staff doing primarily secretarial (rather than typing) work spent only about one-third of their time typing. One-third of all typing was thought to be of the sort that could lead to significant word processing productivity improvements, such as standard letters and long documents requiring frequent changes. It was estimated that about one-tenth of a secretary's time could be made more productive by word processors. Demand for word processors should eventually reach about one-tenth of the number of typewriters in use in 1980.

Although there is a fast-growing demand for word processors, by the early 1980s there was still relatively limited experience of their use over a long period of time. Some organisations installed their first word processor in the late 1970s but usually only in ones and twos and often with equipment from the cheaper end of the market. Very few examples of other advanced forms of text processing equipment were found in the CSA study. Most of the five-year recommendations were based on equipment that was already available in proven and reliable versions in 1980 rather than on more advanced systems which were in a prototype stage or still a gleam in the eye of research workers.

All organisations involved in electronic office projects should expect to go through a long 'learning curve' because it can take many years for the individuals in the organisation to gain solid experience in the nature and potential of the technology and how it affects the unique working environment of that organisation. A slow and steady start at the beginning will become a springboard for more adventurous systems.

It may be tempting for an organisation to sit back and wait for the technology and market to stabilise before introducing new equipment; after all, it is probable that tomorrow there will be new equipment and services at a lower price and with more capabilities. In most cases, however, to wait would be foolish. It is better to use today's systems to make immediate savings and at the same time to gain experience and start the journey up the learning curve. Provided the cost/benefit equation has been evaluated correctly, the investment will be a profitable one in its own right and will put the organisation in a better position to, eventually, exploit more advanced systems. The needs of the organisation should dominate the criteria used for selecting equipment, not just the technological reality or promised potential.

How to organise staff resources

The CSA recommendations covered many organisational aspects as well as equipment requirements. Administrative support groups were recommended for four organisations which would involve several managers sharing one or more secretaries and/or more than one secretary sharing the same equipment. The recommendations which were made in a few cases for the introduction of executive workstations for use by managers and professionals are likely to involve considerable changes in work procedures. It was also considered important for all organisations to have some form of information management coordination as a central responsibility in order to create and monitor guidelines and standards relating to the introduction of electronic information technology systems.

The creation of administrative support groups was recommended as the most likely way to achieve improved productivity. Yet just over half the typists and secretaries in the organisations studied worked alone rather than within a group of secretaries and typists and almost 20% served only one executive. In this situation, word processors could be justified only if the work is regrouped, if the chosen word processor is so cheap that it can be justified on low utilisation and small increases in productivity or if the justification is based on factors beyond typing productivity (such as using word processors as part of an electronic mail service.)

Regrouping work into administrative support units was recommended to four organisations as a way of reducing staff but one of them has a union agreement which stipulated that the introduction of new technology will not mean loss of jobs. Staff reductions were recommended to seven organisations in the medium term.

The sources of typing work and mail

In moving to more advanced systems, an important finding was that in six organisations over 60% of typed work originated in handwritten form. Overall, more than half of all typing originated as handwriting; the source of other typing was printed or typed material, audio or shorthand (in decreasing order of volume.) This suggests that workstations used by professional and managerial staff will eventually have to accommodate using handwriting as an automatic means of inputting information directly to a computer-based system if executives are to find them 'natural'. Otherwise they will have to learn and accept keyboard skills. It also puts into perspective claims that voice input is of prime importance for executive workstations; while voice recognition is undeniably useful in some circumstances, other means of input can be of more importance.

Companies with the highest proportions of audio and shorthand work were those with the highest proportion of staff working for a single boss and, therefore, most liable to change if administrative support groups were introduced. A high level of audio and shorthand did not seem to be a significant ingredient in the productivity of existing systems.

The importance of handwriting as an input technique was also highlighted in the analysis of mail. Although typewritten material accounted for over 60% of all mail, handwritten information was involved in a sizeable 30% of mail, with printed matter only 8% and drawings 1%.

In larger organisations, about three-quarters of mail received by staff was generated by other employees in the organisation. The overall average, including the smaller companies, was just under 50%. Thus organisations have a high degree of control over the form in which mail is sent.

Nearly half the mail is for information only and much of it is stored; paper files were estimated to be growing at about three items per person per day. Storage techniques are therefore a key element in office efficiency.

Evaluation of costs and equipment investment

Salaries were the largest components of office costs found in the survey, with professional, executive and senior management salaries representing well over half the staff costs. On average, secretarial and typing staff represented about 15% of costs but this is higher than the national average of about 10% because the organisations in the survey have a particularly high concentration of paper-based work.

Investment in office equipment, however, was low. A detailed evaluation of one company found an annual investment of £270 per person per year, which is far lower than the average for manufacturing and agricultural workers.

In general, office equipment represented less than 10% of office costs. Although the capital cost of equipment over a period of up to three years recommended in the study was £750 per person, this would still leave office work with a relatively low level of equipment investment.

In most cases, organisations were recommended to obtain 'off the peg' systems of proven reliability which avoid the need for organisations to undertake extensive development of software for their individual needs. There was, however, some requirement for specially skilled staff within the organisation to 'tailor' bought-in software to meet particular requirements and to provide the *interface* links between the various 'building blocks' that will fit together progressively to form a totally integrated office information service.

Short and medium term recommendations

The aim of the CSA strategic recommendations was to develop a long term strategy for text processing within each organisation. Within this overall framework, phased plans were specified to enable new technology to be introduced into the office in a constructive and coordinated manner. The early stages of the plan were divided into two segments: short term covering about eighteen months and medium term encompassing the first three years or so of the plan.

Word processing dominated the short and medium term recommendations but the selection of equipment was affected by the overall longer term strategy. For example, organisations were recommended to install word processors with some communications capabilities although they would not initially be used to communicate.

In addition to word processors, all organisations with more than one site were recommended to use some form of electronic mail within three years.

Only the most experienced organisations were expected to achieve any large-scale integration of systems within the short and medium term though this aim was usually part of the total strategy.

This gradual, phased approach means that emphasis in the short and medium term was on relatively simple technology like word processors rather than some of the grander more futuristic equipment and services which sometimes play a prominent role in images of a paperless, all-electronic Office of the Future.

The following are some detailed recommendations covering the areas that dominated the three-year plans. Similar levels of development could be expected for any organisation embarking on a text processing project. Further details of equipment are given in later chapters.

Word processors

All organisations were recommended to use word processing for the production of standard letters and all but one for the storing and manipulation of standard paragraphs and for simple information retrieval. Other important word processing applications were in the typing of long drafts or documents with complex formats and tabulations and for enabling lists of information to be analysed. In two cases, word processors were recommended for handling scientific formulae. Table 5.1 below summarises the types of applications recommended in the short and medium term.

Table 5.1 Recommended word processing applications

Type	Number of organisations	
	Short term	*Medium term*
Standard letters	9	10
Standard paragraphs	9	9
Simple information retrieval	4	9
Long drafts	7	8
Forms and complex formats	6	7
Simple list processing	7	7
Tabulated material	6	6
Scientific formulae	1	2

In one organisation it was recommended that the word processing should be undertaken on a large computer which was already used as the main information management service. In all the others, additional word processors were recommended.

Word processor systems fall into two main classes, *stand alone* and *shared resource* (sometimes called *shared logic*).

Stand alone systems have their own processing and storage capabilities and operate independently of other workstations. Shared resource systems allow several keyboard workstations to share some central equipment, such as printers, storage and processing power. Different word processing techniques suit different environments:

- *Stand alone* systems, usually involving personal secretaries, were recommended for small branches and remote sites which could cost justify a single word processor for a relatively limited workload or in small departments or groups which wished to maintain independence.

- *Shared resource* systems were recommended in many circumstances, particularly in larger organisations and departments, such as for head office typing pools or where staff reductions could be obtained by creating administrative support groups. Shared resources also assist in situations where many staff need a common information store (name and address list, company standard paragraphs, etc.) or where the processing or storage is greater than that usually found on stand alone equipment.

- *Computer* systems can also be used for word processing. One organisation was recommended to base their word processing on a large (sometimes called *mainframe*) computer. It processed a great deal of textual information but very little of it in the form of the traditional typing activities. In two other cases, the shared resource systems were based on smaller minicomputers which would also undertake some data processing.

The CSA study recommended more shared resource than stand alone systems. This was because the consultants who made the recommendations were more aware than individual purchasers that shared systems generally offer a better prospect to meet long-term requirements to share information and integrate systems. The study also concentrated more on large organisations than small businesses which would create an in-built bias towards larger shared systems. The average price per workstation can be reduced using shared resource systems.

The first stab at word processing in many organisations came with equipment ranging from automatic typewriters to machines with a small single line strip display.

The CSA study, however, found only one situation in which a word processor without a screen was needed in the short term and none in the medium term. Stand alone equipment was recommended to four organisations in the short term rising to six in the medium term. Seven recommendations were made for shared resource systems in the short term and nine in the medium term, the only exception being the organisation which based its word processing on a mainframe computer. In many cases, the recommendation was for a mixture of shared and stand alone equipment.

Administrative support groups were recommended to four organisations. This involved a variety of possible arrangements. For example, a stand alone word processor could be shared by a number of secretaries; one secretary could use a stand alone system to produce long documents for many managers who also had secretaries using manual or electric typewriters; or two or more secretaries and several managers could have access to a small shared resource system.

The consultants did not recommend that all typing work should be done on word processors because the costs in particular situations were too great and the benefits too small or nebulous.

Text input and output

All ten organisations were advised to use conventional QWERTY-style keyboards to enter text. In addition, one short term and one medium term recommendation was made for the use of equipment which recognises handwriting and converts it into a form suitable for text processing. In the medium term there was also one requirement for voice recognition and one for OCR (optical character recognition) of documents. No recommendations were made for direct Computer Input from Microfilm (CIM) or the Microwriter, a 'portable typewriter' with a unique keyboard, strip window display and storage. Since the study was undertaken, however, the price of the Microwriter has dropped dramatically.

The limited range of alternatives to keyboard input do not reflect a lack of need for other forms of input but the equipment on the marketplace at the time was unsuitable because of inadequate capabilities and/or high costs. Such price/performance constraints are likely to be met whenever plans are being drawn up, although more equipment will become justifiable as time goes by because of the relentless march of technological progress.

A similar pattern exists with text output. Two techniques dominate in this area, printers and screens. A host of other equipment is also jostling for a place in the action but in the study other techniques were confined to just a few medium term recommendations. The most important capabilities being sought for some more 'advanced' systems were to have sharper (*high resolution*) graphics to display, for example, signatures, and high speed, high quality printed output.

Computer Output to Microfilm (COM) was thought suitable for two organisations. Photocomposers were also recommended to two companies. One medium term suggestion in favour of an intelligent copier was made to provide a bridge between word processing, conventional photo copiers and facsimile transmission.

Information storage

Generally, recommendations for information storage in the short and medium term were based on equipment widely in use in the late 1970s, such as magnetic floppy and hard disks, microelectronics main memory and microfilm for archival information. The cases studied, however, highlighted the urgent need for cheaper forms of storage to open the way to a wider range of cost justifiable applications including 'electronic filing cabinets' to handle correspondence and document recording on a major scale.

The trends are towards main memory and mass storage which have less electromechanical parts in order to increase reliability and provide quick access at a price which compares with paper and filing cabinets.

This means that techniques like magnetic tape which have relatively slow access times and a large electromechanical content are eventually likely to be phased out. Although microfilm provides a cheap form of storage for archival purposes it does not lend itself readily to computer processing and will eventually be replaced by electronic mass storage devices when they become competitive in terms of cost of storage per item.

In addition to continuing improvement in silicon-chip based main memory (see Figure 2.3 page 24), another form of 'solid state' (that is, without any electromechanical parts) storage which offers increasing potential for low cost, high speed storage is *bubble* memories. Information is stored in them on the surfaces of magnetic 'bubbles' and this technique could store large amounts of information in a small space. Hard disks are also likely to come down in price.

A few long term recommendations tentatively advised organisations to consider some more advanced forms of electronic storage, such as videodisks and holographic memories, provided suitable proven and cost effective systems become available. The videodisk could become an 'electronic filing cabinet'.

Text information retrieval

Computer systems have been widely used for information retrieval services and there are many public and private *databases* of computerised information; the Prestel viewdata service is one example. This kind of database information retrieval has generally relied on largish computers and complex software to organise and manage the database and retrieve information as required. Information retrieval capabilities on word processors are still relatively simple.

Nine organisations were advised to use simple text retrieval facilities on word processors in the medium term plan. Over the same period, there were three recommendations for using the organisations own mini- or mainframe computer for more powerful information retrieval capabilities. Internal viewdata, public Prestel and a link to an external computerised database were also recommended to a few organisations.

The prospect of making large volumes of text prepared on a word processor available to many people (at no extra word processing cost) influenced at least one decision about which locations should be provided with word processors, with priority being given to groups producing information that might need to be looked up widely or frequently. Access to such shared text could, for example, help in handling telephone queries more efficiently.

The effectiveness of information retrieval operations depends primarily on the quality of software that controls this activity. Four organisations were recommended to develop their own special information retrieval software, for example, to link a word processor to an internal viewdata service.

Electronic mail

An analysis of the case studies fully supports previous experience that communications is a prime area of development for text processing. The two most immediate electronic mail applications in the majority of cases were communicating word processors and the linking of word and data processing systems.

Nine organisations were advised to link stand alone or shared resource word processors. Seven organisations which had more than one office site were recommended to implement communicating word processor systems, four of them in the short term. In general, however, recommendations for electronic mail fell primarily into the medium term requirements, with about three times as many system recommendations in this period than in the short term.

Although only a total of four recommendations were made for the use of computers in electronic mail systems, the integration of word and data processing is likely to grow. Some word processing equipment recommended had a little data processing capability, such as simple information retrieval, while some data processing computers also had a word processing option. The main stumbling block limiting the greater use of mixed word and data electronic mail is the difficulty of linking them together in a cheap, flexible and painless manner. Such links are feasible, however, given sufficient investment in skilled resources and equipment to develop networks that allow the smooth flow of information between incompatible equipment from different suppliers. The availability of *local networks* within a building provides a cost effective means of interlinking workstations relatively easily.

Another obstacle to progress in electronic mail was the inadequacy of available software to allow simple and efficient methods of managing electronic mail other than just through the direct transmission of information. For example, software is needed for systems which store information before it is forwarded to the addressee when convenient to the recipient of the mail (known as *store and forward* systems). Shortages of experienced managers to initiate and direct projects and skilled technicians to implement them are also holding back a speedier growth of electronic mail.

Although the medium to long term benefits of electronic mail were thought to be strong, it was not easy to cost justify such systems, particularly in the short term. In some cases, for example, vans were travelling between sites carrying goods so the transmission of internal mail in physical form was achieved at little cost. Informal internal communications, such as tea ladies, were also felt to be effective by some. Many managers did, however, look to electronic mail to improve their personal effectiveness by, for example, transmitting documents between sites in minutes rather than a day or more; reports which need to be prepared at one site and checked or amended by staff at other locations can also be produced more quickly with electronic mail.

Two of the oldest forms of electronic mail, facsimile and telex, did not feature significantly in the recommendations. Although eight organisations already had fax links between sites, no new fax facilities were recommended. The two smaller organisations did not have fax and were not advised to get it; one of these had in fact discontinued use of a fax system earlier as it had not been cost effective.

The two companies which were already major telex users were thought to be ready to link telex with word processors (known as *teletex*), which required software that could route a telex message to a word processor for storage and presentation on the screen. The growing international provision of public teletex networks is easing this software problem.

Internal viewdata/teletext systems recommended for information dissemination were advised in two organisations. Shared resource word processor systems were not regarded as electronic mail although, of course, more than one workstation can obtain information which is in the shared storage.

The integrated electronic office

The concept of an integrated office information system is one which was attractive to all the organisations. Such a system would give full information and adequate electronic processing assistance to help all office workers — managers, professionals, administrators, secretaries, typists, clerks. Such an integrated system has three main components: workstations, shared information storage and a telecommunications network. Such systems would enable tasks to be carried out using a variety of information media, for example, text could be edited, data entered into a computer, information retrieved from a central computer database or from an electronic filing system, messages sent by telex or communicating word processors and links made to an outside network like Prestel.

The CSA study found that integration on a significant scale is still in its infancy in the UK, although some relatively simple first steps were recommended in the majority of organisations. Generally, however, the introduction of services such as word processing, data processing and viewdata systems are developing in their own way without, as yet, a great deal of intercommunication.

Actual recommendations for integration varied from one medium term project for a local 'ring' digital network for a variety of workstations and storage facilities to linking a stand alone workstation to a single shared information base.

A mix of word processing and computing interaction was recommended for two organisations and two others were advised to go along a more text processing oriented route with word processors, central text storage and text communications.

Generally, however, integration plans were based on a fairly gradual moving together of independent systems installed and cost justified in their own rights. The main factor which stimulated progress towards integration was the identification of tasks that would increase the effectiveness of managers and staff other than secretaries and typists.

Products and systems for non-secretarial and typing staff are not emerging onto the market in a cost effective form as quickly as, say, word processors and computers. Equipment incompatibilities which hold back electronic mail developments also impede a quicker integration of electronic office systems.

Costs of short and medium term recommendations

The total expenditure for the detailed systems proposed for the first two phases was £5.3 million, which averaged at under £750 per employee where all levels of staff are taken into account, whether or not equipment was recommended for them specifically. Individual values for organisations varied between £120,000 and £2 million, with systems ranging from comparatively simple word processors to complex highly sophisticated text retrieval and dissemination systems.

An average value of costs is not meaningful as requirements differ so greatly between organisations. The most extensive and highest costing recommendations were in organisations where a high level of experience with existing text processing systems had given management and staff confidence to select more advance systems, such as those aimed at management and professional staff which is less easy to cost justify in simple productivity terms than equipment directed at pure typing tasks.

Although cost/benefit thresholds are continually changing as prices of equipment fall and staff costs rise, it does not follow automatically that equipment should be obtained just because it can carry out a particular task. On the other hand, some equipment should be introduced even if the cost/benefit evaluation is not clearcut because it may bring qualitative rather than quantitative improvements.

Longer term recommendations

Moving beyond the three-year horizon, the recommendations were more broad brush, with few specific equipment specifications. There were three key themes which dominated longer term strategies:

- *Consolidation* of short and medium term proposals and extension to other areas of the organisation.

- *Management productivity* to be improved by introducing systems which would be of more direct benefit to senior staff, such as executive workstations which may include a limited command voice input, direct handwriting input, special function keys and other facilities which move away from the simple QWERTY-style keyboard which many managers are unwilling to use on a regular basis.

- *Integration* of systems developed during short and medium term plans into more comprehensive services, with a quickening growth in electronic mail and other communications services.

Among specific long term recommendations was a system which used synthesised voice to respond directly to voice requests over a private automatic telephone exchange service. Another which specified how an information storage and retrieval system could handle microfilm records of handwritten letters from the public linked to computerised records of action taken in response to the letters.

Software and applications needs

The majority of word processors and other text processing systems are not designed to be programmed by the user. This is a significant difference from computers used for traditional data processing, where a great deal of the software has been written by the organisation using the computer or by external software suppliers who were commissioned to develop specific tailor-made software systems.

Although there is likely to be a growing demand for more flexible and powerful user-programming capabilities, there is unlikely to be an equivalent demand in the office for skilled programming staff as was developed in data processing centres. The degree of variety of service required by office users will be met through the incorporation of software into the basic elements that comprise a system or by the development of 'off the peg' business applications software products that can be used directly or with a minimal amount of custom tailoring.

Three main types of software were identified in the CSA study: business aids to assist management operate more effectively; functional aids to help specialist departments; and software elements built into products.

Business aids

Software can help to organise the manager's own time, improve services offered to clients and enable the manager's work to be performed more effectively. Such systems could include:

- *Personal diary management* which automates activities such as keeping a diary of appointments, planning of meetings, etc. Using a desk-top computer, this type of system could prompt the manager on each day's or week's priorities and might eventually be linked to other individual's workstations so that the computerised diary can be scanned by other staff in the organisation to find out when the manager might be free.

- *Marketing aids* such as being able to store large amounts of market intelligence and client profile information which can be analysed, sorted and retrieved to provide guidance. Overall managing of marketing operations can also be assisted.

- *Mailing* assistance by, for example, automatically producing standard letters plus names and addresses on envelopes, individually typed by a word processor. Automatically produced mail shots could be made selectively only to those people in the client profile file whose records are identified with specified characteristics, such as being located in a particular area.

- *Internal communication* can be made more efficient because longer documents can be produced and disseminated quickly and all other information can be circulated speedily over an electronic mail service. If an electronic 'Mailbox' store and forward feature is available, managers can be sure that a message needs to be sent only once when trying to contact someone, rather than leaving a message that might go astray or be forgotten.

- *Corporate databases* in larger organisations can contain extensive information that can be accessed from all over the organisation via information retrieval workstations or viewdata or teletext sets; software for such databases may be so large it has to run on mainframe computers.

Specialist functional software

Software that could automate common business functions were frequently required. Such software could be used to run text processing tasks on a word processor or data processing activities on a local word processor rather than a remote centralised computer. The specialist functions include:

- *Accounting* tasks which have been the bread and butter of commercial data processing, such as sales ledger and bought ledger. In a large company, word processors in particular departments could get information for accounts processing locally and from a central information base or accounting function. Software to process this information could be stored on a central computer and brought into the local word processor when required to run the required accounts routine. One company in the study was recommended to use a word processor to do the payroll for temporary staff and casual labour, a task which was inappropriate for processing by centralised payrolls on large computers.

- *Library* systems involving word processors and photo-typesetting were recommended which were more limited than the type of library services widely available on large computers for bigger libraries. Word processors can assist book ordering, cataloguing, loans and searching. For example, they can produce standard letters for orders and use electronically stored information to print catalogue cards and accession lists. They can monitor book loans and trigger the automatic production of standard reminder letters if the books are overdue and provide information retrieval services.

- *Membership subscriptions.* One type of list processing software for word processors which was recommended was to handle membership records. For example, information could be retrieved relating to a group of members with particular characteristics, such as being over a certain age, or standard letters could be typed to selected groups of members. Linked with the membership management software could be a system that handles membership accounts and produces invoices, reminders, with an automatic prompting to send standard letters when subscriptions are due.

Software elements to assist particular hardware

Some software requirements relate to enhancing particular hardware services either as a program incorporated in the hardware product or as an interface device or service that helps to interlink different equipment. The survey identified the following functions that depend on suitable software elements and which were of particular importance:

- *Telex* interfaces with word processors.

- *OCR* which interprets the meaning of information that is input via the hardware reading mechanism.

- *Viewdata* interface software which enables word processors to be used to update information in private viewdata systems or the public Prestel service.

- *Phototypesetting* output direct from a word processor which depends on software to handle the formatting and layout tasks.

- *Archiving* information onto microfilm or optical disks with software to transfer information from word processors in an appropriate form and to create suitable access paths via indexes for retrieving that information.

- *Facsimile* did not feature significantly in the study recommendations but there were some requirements for interfaces to enable output from a word processor to be digitised and transmitted to a remote fax receiver.

- *Information retrieval* software needed to help search through files on the basis of key identifiers specified by the user and also to interface with external information services which often have their own search techniques.

A solid basis for future planning

The picture that emerged from the CSA strategic studies provides a realistic counterbalance to the more extravagant claims that new technology will suddenly alter the nature of office work. Although electronic information technology equipment was recommended to all organisations, the innovations were placed within the context of gradual organisational evolution.

The recommendations covered the period to the mid-1980s and beyond. The strategic framework, however, is applicable whenever an organisation is preparing for significant technological innovation. The main actions should be:

- *Establish* a central policy under senior management direction.
- *Create* a long-term strategy.
- *Develop* detailed phased plans that start simply but always conform to longer term objectives and systems development standards.
- *Allow* for a lengthy period where managers and staff gain experience and understanding of relatively simple technology before moving to any more ambitious schemes.
- *Give priority* to human and organisational factors.
- *Do not* implement new technology just for the sake of it; any new technological step must be justified in terms of overall organisational objectives and specific application needs.

Over time, some of the equipment which was regarded as unjustifiable at the time of the study will become reliable, cost effective parts of many systems. The main stumbling blocks, however, are not so much the availability of independent equipment with particular capabilities but with linking together and integrating existing systems, for example, into an electronic mail network that allows a variety of types of equipment to communicate with each other easily, flexibly and cheaply.

The target is integration

The information in this chapter provides a solid basis for keeping technological developments in perspective and for acting as a guide to developing detailed plans for a specific organisation.

Word processing and electronic mail are competing for the claim of being the 'heart' of the electronic office. These are extensions of the two most important items of traditional office equipment, the typewriter and telephone. In the majority of organisations studied by the CSA, the word processor was recommended as the cornerstone of integrated office developments.

There is also, however, a hypothesis that office systems will be built around the telephone and other communications systems. For example, the Delta system from the US Delphi Corporation (which is marketed in the UK by Nexos Office Systems) was first used as the basis of a telephone answering service in the US. With suitable software modifications, it can be used as an office information system.

Whatever the starting point, the end objective is universally agreed: to create an integrated electronic information system to handle text, data, voice, images, handwriting and other natural forms of human communication.

The next two chapters examine the nature and benefits of word processors and electronic mail in more detail.

6 Focusing on the word processor

The driving force to the electronic office

Word processors can produce dramatic increases in typing productivity and can, therefore, easily be cost justified. They can also aid management effectiveness and improve the speed and quality with which letters and documents are produced. At the same time, a word processor can be the kernel of wider text and information processing activities. As a workstation in an electronic mail network or with sophisticated software that enables it to carry out complex data processing and information retrieval tasks, the word processor can become a powerful computing facility in its own right.

Word processors, more than anything else, are therefore likely to be the driving force towards an electronic office because they can provide immediate returns on investment and fulfil long term strategic requirements.

The simplest word processor is an automatic typewriter. This is similar to a conventional typewriter but has some limited storage capacity and can perform simple text editing functions like inserting or deleting text. More advanced equipment have video screens which can be used to display text and powerful software that can perform complex editing tasks such as moving, copying and other manipulations of blocks or columns of text.

Computers can also be used to perform word processing but the most popular forms of word processing are systems where text processing is the prime priority rather than having typing activities as subsidiary to other data processing operations.

Where word processors are particularly useful

Benefits of using word processors vary considerably depending on the nature of the typing and output being undertaken, the equipment used and the performance measurement techniques specified. In the right circumstances, productivity increases of about 200% could be regarded as a feasible target. As a rough guide, increases in productivity of 60% to 150% can be expected for stand alone screen based word processors and 100% to 250% for shared logic systems.

These figures are based purely on typing activities. In addition there are other benefits which are difficult to quantify precisely, such as saving executive time spent checking and reading documents that go through a number of typed drafts. Word processors can also improve the quality of presentation of documents and speed up the time to get material typed.

Some of the improvements in productivity can be translated into staff savings. In some organisations, the more satisfying work environment that can be created by a well-implemented word processing system can also cut staff turnover. One organisation has recorded a reduction in staff turnover from 30% a year to 2.5% following the introduction of a word processor.

Organisations are generally reticent about publishing cost savings from word processors. In addition to the kinds of typing productivity increases already identified, however, there is ample practical evidence of the total benefits that can accrue from the use of word processors and related technologies. The Bank of America, for example, estimated a saving of about £3 million in 1978 attributable to word processing. One UK organisation has saved almost £100,000 in one location and an insurance company has increased its typing workload by over 500% in four years and at the same time decreased staffing by almost 24%. One organisation which was recommended text processing equipment during the CSA study expected to increase sales by £30 million a year due to the services offered by the new system.

Such benefits are impressive and tempting. Yet there is no guarantee that the introduction of word processors will automatically produce such dramatic benefits. In fact, there are many cases where word processors have led to organisational disruptions, staff dissatisfaction, increased error rates by typists and poor efficiency.

These failures have usually been caused more by management and organisational inadequacies than by problems with the equipment itself. Inadequate training of management and staff in how to change their methods of work to exploit the capabilities of the technology has often led to grossly inefficient use of the installed equipment. The creation of word processor centres, which was fashionable during the first wave of word processing activities in the 1970s, frequently gave superficial typing productivity improvements but impaired the overall efficiency of the organisation because managers and professional staff were often deprived of the essential administrative and clerical support. This caused managers to waste a great deal of time, for example, doing their own photo copying. The administrative support groups recommended by CSA consultants during the text processing strategic studies aim to ensure that equipment resources are shared and coordinated efficiently while at the same time providing the personal assistance needed to help senior staff in becoming more productive.

Not all typing activities justify the use of word processors. Short memos and letters which require little correction, for example, usually yield little productivity or quality benefits from the application of word processors. The tasks that are to be handled on word processors should, therefore, be carefully selected to direct word processing power where it is likely to be most useful.

The following workloads and environments should be considered as high priority for the application of word processors:

- long drafts with frequent amendments and retyping before a final version is approved (annual reports, speeches, conference papers, management plans, etc.)

- a high proportion of standard letters, each of which requires slight variations, such as name and address

- many frequently produced documents made up from standard paragraphs (legal and insurance work, personnel staff contracts, etc.)

- work involving forms or presentation in complex formats or with many tabulations

- a significant number of documents containing scientific formulae

- a large amount of mail shot activity which involves, for example, merging a list of names and addresses with a standard letter

- the frequent production of long reports and other documents where presentation is important and timescales are limited

- in typing pools where a shared resource system could cut down on a great deal of paper handling

As part of an electronic mail service, the word processor could enable documents and messages to be transmitted more efficiently and quickly than alternative means, such as fax or telex. The word processor could be used to avoid duplication of form filling effort by entering information directly to the computer using a document format presented on the screen which is 'filled in' by the word processing operator as necessary. Word processors could also interact with data processing systems to produce, for example, purchase orders and invoices and can be linked to an output device such as a typesetter.

Types of word processors on the market

The term 'word processor' was coined when magnetic storage was first used by IBM in the 1960s with an automatic typewriter. In 1973, US companies like Lexitron and Vydec introduced word processors with screens and software which could perform complex editing tasks on the stored information. A year later, in the UK, Unilever and Logica produced the first shared logic system that enabled many workstations with screens to access the same storage and processing capabilities.

Text and word processing packages are also available on mainframe, mini and small business computers. Such systems are more oriented to having data processing as their main priority and have not taken a significant share of the market for word processing aimed at helping mainly with office typing workloads. There are also 'hybrid' systems marketed as combined word processor/data processing packages. The vast majority of systems recommended during the CSA study were, in the short to medium term at least, primarily for stand alone and shared logic word processors.

In 1981 there were well over 100 models of word processors on the market in the UK from over 50 suppliers. The number of word processing workstations was rapidly approaching 20,000, with well over ½ million units in use throughout the world, mainly in the US. The capabilities of the different machines vary considerably but can be grouped into five main categories:

- *Basic automatic typewriter* which looks like a conventional typewriter but stores information on magnetic card, cassette or diskette, although text is also committed to paper as in a manual typewriter. When performing one of its limited range of editing facilities, like insertion or deletion, the typist replays the stored information and has to wait while the text is automatically retyped up to the point where text is to be deleted or inserted. As text is always typed onto paper, if an error is made then the whole document or page has to be reprinted.

- *Advanced automatic typewriter* usually has duplicate storage and may be capable of communicating with a computer. Otherwise has similar disadvantages and limitations to the basic automatic typewriter.

- *Word processor with strip window display* which provides a major advantage over automatic typewriters through the provision of a limited, single line display (about 20 characters) which acts as a visible 'buffer' between making the keystroke and the text appearing on paper. This can improve typing productivity significantly because relatively trivial typing mistakes which form a significant part of all typing errors, such as transpositions, can be corrected before the text is committed to paper.

- *Stand alone screen word processor* which consists of a screen, similar to a TV monitor, a keyboard to type input, a processor and software, and a letter-quality printer. The provision of a screen (which typically contains about 20 to 30 lines of text or more) enables a wide range of sophisticated editing functions to be performed manipulating blocks of text. The editing capabilities are provided primarily by software. The potential of a particular machine to keep abreast with new developments therefore depends more on the availability of suitable software over a long period than any particular hardware capability. Printing takes place separately from editing so that it is possible for the typist to be preparing and editing a new document while the previously edited document is being printed. In theory these machines are much easier to use than the automatic typewriter. In practice, however, some are far more acceptable to operators than others. The ease of operation and the consideration given to operators' comfort are important criteria to look for when selecting equipment.

- *Shared resources/shared logic screen systems* which can have many keyboard/screen workstations linked to common storage (usually on hard disks), processing power and printers. Such systems provide savings on resources and greater flexibility because, for example, fewer printers are required and one printer could be a high speed, relatively low print quality unit for producing drafts. The provision of central storage means typists do not have to be responsible for handling a set of magnetic cards, cassettes or diskettes. Shared resource systems also relieve the typist of responsibility for paper handling, which can be coordinated by a print controller. For long and urgent documents, many typists can work in parallel using a shared logic system with their output merged into a single consistent document using editing capabilities. These types of facilities give shared resource systems a productivity edge over stand alone systems. To help in the effective coordination of the shared resources, facilities should be available for use by a word processing supervisor to control, for example, the storage and deletion of documents.

Management, staff and organisational objectives

When selecting word processing systems, the equipment should not only be capable of providing specific capabilities in the most cost effective manner but should also fit into the overall management strategy. It is not necessarily the cheapest equipment or the one with the most advanced facilities which is the best choice. Management should be satisfied that the supplier has a long-term commitment in hardware, software and maintenance for a variety of electronic office information technologies and is commercially viable.

Four broad principles should govern the selection of equipment:

- *Select* a company with a stable committed future in word processing and a broad interest in the 'Office of the Future'. It is important to be sure that enhancements will be made continuously.

- *Satisfy* the organisation's *major* needs rather than looking for equipment that does everything but may not be good for the applications that dominate the organisation's operations.

- *Standardise* on one type of word processor (or perhaps two if the organisation is sufficiently large.) It is important to be able to transfer work and staff from one machine to another. When the organisation starts thinking of moving to an integrated system, the problem of linking equipment is far greater if several types of equipment is in use.

- *Expect* at least simple communications at some time between word processors and other equipment.

The introduction of word processors could also trigger organisational changes. These should be carefully thought through before investing heavily in equipment, as the nature of the organisational changes and the resultant training could influence system selection. For example, there is likely to be a move away from personal secretaries towards integrated groups of administrative support staff. This will make shared resource systems more attractive and will create a demand for supervisory facilities to coordinate word processing work.

It is also likely that organisations will move progressively to greater coordination of information management functions such as data processing, typing, communications and libraries, so the word processing plans need to be coordinated with other managers or, if it has been established, a corporate information management unit.

The following checklists summarise important capabilities that may be of significance to management and other levels of staff when selecting equipment:

General systems capabilities

- Is the system oriented towards handling complete documents or is it page oriented, with re-pagination decisions involved every time page boundaries are crossed?

- How long does it take to initiate the running of the system and how long for individual staff to 'sign on' (*log on*)?

- Does the system have a spelling correction capability? If so, does it just highlight incorrect words or attempt to correct them? What size dictionary is checked and how many special words can be inserted by the user to increase the standard dictionary's vocabulary?

- Are other 'dictionary' capabilities available, such as translations of common words or hyphenation rules? Is hyphenation at the end of lines completely under operator control or semi-automatic?

- Are explicit operator commands needed to delete documents or can they be done automatically according to pre-programmed criteria? Can documents that are to be deleted be placed in temporary storage in case of an error?

- How is the time taken for the system to respond to an operator input (the *response time*) affected by the number of workstations connected to the systems? If there is a serious drop in response time, does it affect all operations from these workstations?

Management and supervisory facilities

- The ability to design special forms, possibly using the screen, for storage in the system so that the format can be recalled by the typist who then just 'fills in the blanks' from the keyboard.

- Provision of statistics, operator performance, print utilisation, etc.

- Availability of security checks, such as passwords, to provide multiple levels of automatic security to prevent unwarranted access to particular categories of information.

- Housekeeping and archiving software, particularly on shared resource systems, to give the supervisor assistance in optimising the use of information storage (see page 79).

- A document *directory* containing information such as the name, date and length of documents helps to keep track of information in the system.

- When faults occur in the system, information should be available to identify and diagnose the fault in hardware and/or software.

- Provision of hardware and software performance statistics.

Training and documentation

When a system is introduced, all levels of staff are likely to need some form of training to understand how to prepare work to optimise the effectiveness of the systems as well as operator training. Training instruction on some systems is provided directly by software which takes the operator through the facilities step by step, under the control of the computer. Good, clear, easily understood documentation can also ease training requirements. Suppliers should be questioned on the scope, cost and quality of the service offered, such as:

- Training aids (films, brochures) for management and professional staff.

- Word processor supervisor, typist and other operator training.

- User familiarisation courses and assistance.

- The time taken to train typists and supervisors to a level where they can be confident of obtaining useful levels of output.

- Whether new users of the equipment must be trained within the organisation or will the supplier be involved each time.

Equipment interconnection capabilities

As the word processor is likely to form part of a wider system eventually, questions should be asked about the interconnection facilities. It should be determined whether such interconnection is standard with the system, made by the purchase of a special interface option or dependent on special development by the supplier and/or the organisation. Interconnection capabilities that should be investigated include:

- Direct links to telex, OCR Optical Character Recognition input, other word processors or computers. Particular equipment models should be specified when making enquiries.

- Direct or indirect information transfer via paper tape, magnetic tape or disks. Specific details of each medium needs to be investigated, such as tape width, coding conventions for holding the information, disk storage density.

- Can communication with other equipment take place simultaneously with inputting or editing or must normal working cease while communication takes place?

- The communication *protocols* and standards which are supported, such as the IBM 2780 Communications protocol or the international standards such as X25 and HDLC (see Appendix 1).

- Functional characteristics such as the range of commands available, the ability to communicate interactively (with immediate response to enquiries) or being able to cope with particular information coding standards, such as ASCII or EBCDIC (see Appendix 1).

The word processor keyboard

The prime means of entering information to a word processor is a keyboard. This consists of similar alphabetic and numeric (*alphanumeric*) keys as on a conventional typewriter keyboard, plus various control keys and special functions which have specific relevance to word processing.

The most common alphanumeric keyboard layout is known as QWERTY, because these are the first letters in the alphabetic row. This was originally invented in 1873. With the traditional 'typebasket' printing mechanism for manual typewriters, the keys could jam if the typist pressed the keys too quickly. The QWERTY layout slowed down typing and spaced the frequently used keys to prevent jamming.

Although this layout is unnecessary for modern single element print mechanisms, such as the golfball on electric typewriters, the convention has persisted because there are many generations of typists who have been trained for QWERTY keyboards and there is an enormous investment in equipment with this layout. This is a good example of how technological changes can be tempered by non-technological factors, such as staff expertise, existing investment and historical traditions.

For European languages other than English, the AZERTY layout (these letters are in the equivalent positions to QWERTY) is frequently used, and also includes an appropriate range of accents and diacritics.

In 1932, Dr August Dvorak, Professor of Education and Director of Research at Washington University, introduced a simplified keyboard based on the analysis of the most commonly occurring words and letter combinations. He claimed that it would accelerate typing speeds by 35%. Despite the fact that it is comparatively easy to produce on standard production runs, there has been little demand for Dvorak's keyboard.

More recently, Lillian Malt and Stephen Hobday in the UK

have designed the Maltron keyboard which, like Dvorak's, positions the letters to maximise speed. The Maltron is also shaped to match different finger lengths and the keys are located in a way to ease two-handed typing. The company PCD has incorporated the Maltron keyboard into a portable word processor on a trolley which, it is claimed, could be wheeled into a manager's office and used to type as the manager dictates the information. Keying speeds with the Maltron are claimed to be up to 40% faster than with QWERTY.

Control keys to perform specific word processing functions also form part of the keyboard. It is important that these keys are clearly distinguished from the rest of the alphanumeric keys and are clearly marked for the functions they perform.

An important function of the keyboard is to control the pointer on the screen (called a *cursor*) which indicates the place in the text that is currently being worked on. Four or five keys are provided for cursor control for movement up, down, left, right and, on some machines, return to a 'home' starting point.

The cursor can also be controlled by devices called *cats* and *mice* which translate hand movements into signals to move the cursor. Cats were developed by Xerox Corporation. The operator 'strokes' the cat by moving a finger in the desired direction over a touch-sensitive pad. A mouse is a unit which is moved across the desk in front of the screen. This triggers the cursor to move in the same direction. The mouse unit might contain its own keys which can be used to indicate actions to be taken once the cursor has been positioned.

It is also possible to link a portable keyboard to a word processing system over telecommunications links, for example to enable an insurance salesman to transmit details of a policy. Such a keyboard could form part of a portable 'terminal' that could also include its own output device, such as a printer, and its own intelligence and storage (see Chapter 9 for more details on text entry devices.)

Summary of desirable keyboard qualities

Because of the importance of the keyboard in the word processing job function, it is important that its ergonomic design is examined to ensure that it will provide a comfortable and efficient working environment. The keyboard should be capable of fitting into a work surface which provides sufficient room and flexibility to enable the typist to sit comfortably and to view source documents in a way that does not involve too much head or eye movement. Ergonomic factors dominate the following summary of keyboard facilities that should be looked for when selecting a word processor:

- Overall simplicity and logicality of keyboard layout to aid efficiency and the typist's comfort.

- Clearly defined functions keys, preferably separated from the QWERTY alphabetic keys and a different colour to the other keys.

- A keyboard unit separately moveable from the display screen. The unit should be slim enough to enable it to be positioned at a convenient height and should be set at an angle that is comfortable for the typist.

- The keys should have a good 'feel' when pressed in operation.

- A separate numeric keypad can be useful for some applications, possibly as part of the keyboard.

- The cursor movement for screen systems should be repeatable by simply holding down the appropriate cursor control key.

- A clearly marked shift-lock indicator.

- A repeat key to enable the same character to be repeated while both it and the repeat key are pressed at the same time.

Information storage for the word processor

Magnetic media are the most popular form of storage with word processors. The magnetic card from IBM was the first word processor storage medium. Now there are various forms of magnetic tape and disks in addition to the card. More advanced, compact and speedy electronic methods, such as videodisks, offer great potential when they become available on the market as proven, cost effective products.

The magnetic card is just over 3 inches by 7 inches and is coated with a magnetic film that can hold about one page of typescript, of about 50 lines each with up to 100 characters.

A range of magnetic tape storage media is available. Tape widths vary from ¼ inch to one inch. Tape feeds can be from open reels, closed cassettes and cartridges or endless tape loops, containing from one to about 14 tracks. A typical cassette can hold about 80,000 characters. Some manufacturers provide their own types of tape, such as the IBM Sprocketed tape which can store from 7,000 to 28,000 characters depending on length and the Olivetti one inch loop that can store about 250,000 characters.

Magnetic disks are either flexible or rigid. Flexible disks are thin, flat pieces of plastic, usually about 7½ inches in diameter, coated with magnetic film and called floppy disks or diskettes. A floppy disk is held in a protective envelope which remains in place when the disk is rotated on a spindle so that information can be read from or written to it. A full floppy disk has a capacity of about 250,000 characters, equivalent to around 100 pages of text. Double sided and/or dual density disks are also available, as are mini-disks which are about 5.5 inches in diameter and have about half the capacity of a full-sized diskette.

Rigid (or hard) disks are available in a cartridge form and can store from about 1,000 to 5,000 pages. A major advantage of disks over tape is that information can be accessed directly at any point on the disk rather than having to read the tape from beginning to end to find the appropriate position, just as a track on a record disk can be selected directly. Larger hard disks, as used for computer systems, are also available; these are suitable for shared resource systems.

Videodisks could have capacities of 50,000 documents or

more but their use will be limited until software systems have been developed to access information in the disks as well as making the hardware cost effective.

Storage can, of course, be used to hold both programs and other information. Silicon chips are generally used as the main memory to hold programs and information needed to carry out the immediate operations. Bubble memories of around one million bits and more on a chip are becoming an alternative to some storage requirements previously met by magnetic media. (See Chapter 9 for more details on information storage media.)

Summary of storage characteristics

The choice of storage medium will depend on the volume of information to be handled at each workstation or by a shared resource system, the cost of the equipment and the appropriateness of the methods needed to handle the medium. The following are some important factors relating to storage that should be considered when selecting a word processor:

- A true picture should be determined of the number of characters that can be held rather than a theoretical statement of total capacity (which could include areas used for control or other purposes).

- The medium should be easy and safe to handle and should enable security copies to be made and stored at a separate site.

- Some types of information should be regularly copied (*dumped*) from the current storage to backup archival files in case something goes wrong with the operation of the word processor and the state of the information shortly before the break has to be reconstituted; this is known as an *archival* facility.

- The life expectancy of the medium should be checked.

- The medium should be capable of being safely handled by unskilled staff unless it forms part of a larger system where storage librarians may be employed.

- It may be necessary to retrieve items according to a date or reference number, and such a facility should be available in software, or because of the nature of the medium (a cassette or diskette could be dated by hand.) Reference numbers could be generated automatically, manually or a combination of both.

- *Housekeeping* software should be available, for example, to enable supervisors of shared resource systems to allow documents to be added, removed or rearranged on storage media.

- When the system reaches the end of a particular storage unit, it should automatically handle the overflow onto a new unit without inconveniencing the operator or losing information that has already been keyed in.

- Storage can be allocated in a variety of ways. For example, complete tracks or large sectors on a diskette could be assigned to a particular information unit, such as to each document, even if some of that space is left unfilled. Other media allow for small block allocations to optimise the available storage. In conjunction with housekeeping software, greater value for money can be obtained by carefully selecting and using media that can pack more useful information into space rather than judging cruder theoretical figures on total storage potential.

- The time to store or retrieve a one-page A4 document should be compared to the equivalent times for multi-page documents. If there is a significant difference, the reason should be determined. The limitations on document length should be investigated.

The scope of text editing

The editing facilities required by a typist using a word processor are similar in some respects to those required in the newspaper industry. In fact, the term *author* is frequently used in word processing systems to describe the originator of material, usually a manager, professional or other senior staff.

With conventional typing systems, the author faced with a first draft frequently resorts to traditional 'scissors and paste' methods to chop the text around and insert new text, as well as writing in changes in the text and comments to the typist. When the second draft is returned, the whole text often has to be read through again because the changes were so extensive that the typist had to start from scratch and retype the whole documents. New errors can, therefore, be introduced in text that were correct in the first draft.

Editing procedures using a word processor are different and more efficient. The author marks up the draft by indicating where characters, words, sentences, paragraphs, etc., need to be deleted, inserted, moved backwards or forwards, amended or have some other editing process performed on it. This is similar to the technique used by sub-editors and proof readers in publishing.

The first draft text must not be physically cut up and moved about because it exists in the word processor memory in the form of the first draft. When the typist receives the marked up copy of 'proof', the pages of text are displayed on the screen. Using the cursor as a pointer, the typist/word processing operator searches through the text and types in editing commands, such as DELETE or INSERT, when the cursor is pointing to the appropriate character, word, line, paragraph or block of text.

All the editing manipulations of text occur within the computer memory, not on paper. Any first draft text that was correct is left intact although the text around it can be chopped and changed. Once the changes have been made, the revised text can be printed with any format required. For example, if a sentence is deleted from a paragraph, the second draft of that paragraph will have the correct margins and spacings.

This process continues until a final draft is approved. Compared to traditional methods, the final draft is achieved more quickly, with authors spending less time reading, rereading and marking up drafts. Typists need fewer key-strokes to reach the final draft because only the amended text needs to be manipulated and input. Correct text typed in an early draft does not need to be retyped in subsequent drafts.

Although the basic editing functions are similar to those used in setting text for phototypesetting, the word processing functions are generally easier to operate and are far less demanding than the facilities required for printed text formatting when preparing a newspaper page.

More and more editing facilities are being added to word processing software every year, many of which are aimed at specific applications. For example, totals can be automatically checked on tables of figures.

A summary of editing facilities

The following are some of the main editing features that should be sought when evaluating word processors:

- Deleting a character/word/paragraph/string of text/block/page.

- Inserting a character/word/paragraph/string/block/page.

- Reformatting of text after an insert/delete.

- Moving or deleting any size or shape block of text.

- Working in column formats with the ability to move columns to the left or right and to delete columns (known as *columnar* working).

- Transferring text from one document to another.

- Setting margins and tabs automatically; the top line of a screen is often used as a 'tab rack' to indicate tab and margin settings.

- Assembling standard paragraphs into a specific document.

- Centring of headings between margins and tabs automatically.

- Lining up text evenly on right hand margin (*right justification*).

- Merging of lists (say of addresses) with standard text, for example, to carry out a mail shot.

- Searching for a specified string of characters automatically.

- Reformatting of text within new margins.

- Aligning decimal points in tabulated numerical work.

- Providing *wraparound* at the end of a line. This allows text to be typed without concern about reaching the end of a line because the systems will automatically start a word on a new line if there is no room for it on the current line.

- Eliminating and/or flagging *widows*, the term given to a single line left on a new page or a single word on a new line. Widows can be wasteful of space and paper and are generally regarded as unattractive.

- Identifying when the end of page is reached.

- Exchanging a word or string of characters by something different via an automatic search and exchange operation throughout the text (called *global exchange*).

- Allowing for proportional inter-word spacing on output.

- Renumbering pages automatically when the text has been lengthened or shortened when editing a multi-page document (called *pagination*.)

A paperless office

Traditionally paper has been the dominant medium used for presenting text output. Typing, printing and copying equipment have therefore represented the major proportion of investment in office output equipment. With electronic text processing systems, printing plays an important but less singularly dominant role. In addition to printed output, the results of word processor activities can be presented on a screen display or film, as direct output to an electronic mail service, or to control devices such as voice synthesisers.

In the long term, the various alternatives to paper may eventually lead towards the so-called 'paperless office'. This will not be a sudden change because paper remains a highly competitive medium for storing and handling information both in terms of cost and convenience. New printing techniques which provide improved quality at higher speeds will also prolong the life of paper by pushing forward its cost effectiveness. For some purposes, paper is also the most appropriate medium.

The display screen is, however, rapidly becoming a

common means of output in offices, either as part of a word processor, as a Visual Display Unit (VDU) linked to a computer or as a TV set linked to a viewdata or teletext service. A word processor's display plays a role in the input as well as the output process because it provides a visual representation of the typist's key strokes.

The screen can be used as a direct replacement for paper because formats of standard documents can be stored in the system and displayed on the screen when required. The cursor on the screen can be programmed to move automatically to the spaces where information should be filled in and this is done by keying in the information, which is then stored in the system.

Although paper is likely to remain a relatively cheap storage medium for some time, in some circumstances screens are an economical alternative as an output medium. The set-up costs of providing displays and communications services for text oriented electronic mail is usually greater than the cost of providing a printer. Once the displays have been provided, however, the marginal costs of using them are less than the marginal costs of paper printing and distribution. The viability of displays as an output medium in any location will depend on the volume of information going to that location as well as its urgency.

Characteristics of word processor displays

The commonest form of display is a Cathode Ray Tube (CRT) as used for TV monitors. Characters are displayed as a matrix of tiny dots which form the appropriate shape. These dots are created by the CRT which acts as an electron 'gun' focused onto a phosphor-coated inner screen; when the beam strikes the screen, the phosphor point glows. This is sometimes called *raster scan* image generation.

There is also a technique called *vector generation* which produces characters as sets of lines. Flat *plasma* screens composed of tiny neon tubes use vector generation and can provide a full A4 page size display. Flat TV screens are also being developed.

As word processor operators are likely to spend a great deal of time working with a screen, ergonomic factors such as the quality of the image, the lack of screen glare, flexibility in positioning screen, etc., must be given high priority. The health and safety aspects of word processors have been brought to prominence by staff and unions' reports of the possibility that working with VDUs could create eye cataracts, minor facial skin rashes and other complaints.

Although there is much controversy about whether VDUs are the direct cause of such complaints, there is substantial evidence that poor quality displays and cramped, ill-lit and otherwise poor working environment can cause mental and physical distress and lead to ineffective operation. Staff working regularly with screens should be given frequent eye tests to ensure they do not have a condition that could be aggravated by constant viewing of text on display.

Some standards are beginning to emerge on the ergonomics of screen-based systems. The *Visual Display Terminal Manual* by A. Cakir, D.J. Hart and T.F.M. Stewart (John Wiley & Sons) provides a detailed look at these aspects. Some unions have defined standards as the basis for industrial relation negotiations on the environment in which technological innovation takes place. Management must, therefore, be satisfied about the ergonomic design of the equipment during the evaluation phase. This can be done by talking to staff with operational experience of the system, consulting experts and giving staff a chance to use equipment as part of standard evaluation procedure. Ergonomic considerations extend to the whole unit, not just the screen. Keyboard layout and design is just as important as screen construction.

The following are some characteristics that are applicable to screen displays; the precise requirement for these will depend on the nature of the particular tasks for which the word processor is being used. These factors can be divided into capability and ergonomic features. The capabilities relate to the provision of new or more sophisticated information handling and display functions; ergonomic factors are of prime value in making the word processor operations a more satisfying, comfortable and effective job.

Some advanced display capabilities

- The availability of a multiple colour display as an option; such displays are more expensive than the monochrome or two-tone displays. Although colour TVs can be used to display text, say, for a viewdata system, the quality of character presentation is inadequate for use in a word processing application where the typist is likely to spend a significant amount of time looking at the screen from closer quarters.

- *Scrolling* which is needed because most screens can display only the number of characters that would usually fill about half a page of A4-size text. The screen could be regarded as a window pointing at the document. Vertical scrolling moves the 'window' forwards (down) to text ahead of the current position or back (up to previous text). Horizontal scrolling moves the 'window' to the left or right. Questions should be asked about what types of scrolling are available (forward, backward, left, right), whether sideways scrolling decrease the number of lines displayed; and at what point does text move up the screen.

- The number of characters per line before sideways scrolling begins and the number of lines of text displayed *excluding* tab and control information lines may be of significance.

- The provision of underlining.

- Availability of many character fonts (different type face designs.) Alternative character sets may be selectable by software or hardware switch.

- The availability of brightening and reverse video (say, black on white for particular characters, words or larger blocks of text.)

- Split screens or multiple windows have been developed for some displays to enable more than one page or document to be shown at the same time. The standard form of display can have only one 'frame' at a time to fill up the display area.

- Cursor flexibility and speeds may be of significance. Some systems allow the creation of a 'ghost' cursor which acts as a temporary pointer while some subsidiary action is performed in one part of the screen, leaving the permanent cursor at its original position. Some cursors flash on and off to highlight their position but this can be more distracting than helpful.

- Graphics displays may be required for some tasks rather than just text. This could depend on both software and hardware capabilities.

- Subscripts and superscripts may be needed, particularly for mathematical and scientific work, with characters capable of being displayed a little lower or higher than the main line of text.

- The ability to display text with proportional spacing (where the letter 'm', for example, takes up more space than the letter 'i').

Ergonomic factors

- Maintenance of a stable, 'flicker-free' image. With CRTs, as soon as an image is formed on the phosphor, it begins to fade and so has to be regularly refreshed; instability may occur during refresh or because of some variability in the power supply to the equipment.

- Character images should be sharp, without any 'fuzziness'.

- The screen should be an independent unit, not attached to the keyboard. This gives flexibility in positioning the equipment.

- The screen should be capable of being rotated horizontally and vertically so that it can be positioned at the most convenient angle for the operator.

- The screen and office environment must be chosen to avoid reflections, glare and other possibly disturbing environmental conditions.

- There should be a clear distinction between similar characters, such as X and K, O and Q and S and 5.

- Adequate character size and shape. The following recommendations come from the white collar union APEX's report *Office Technology*:

The Trade Union Response:

Minimum height (at 70cm viewing)	3.1 to 4.2 mm
Maximum height (with 5×7 dot matrix)	4.5 mm
Width to height ratio	3:4 to 4:5
Spacing between words	half character height
Spacing between lines	full character height

If viewing is from other than 70 cm, the character height should be one two-hundredth of the viewing distance.

Printers available with word processors

There are two main types of printers for producing paper output for word processors: *impact* or *non-impact*. Impact printers involve a print head actually touching the paper; a conventional typewriter is a form of impact printing. Non-impact printers form the image by placing ink on the paper without having a character actually hit the paper. Word processors can also provide direct input to phototypesetters. Non-impact techniques like ink jet and laser printers can provide greater flexibility than impact printers.

Impact printers

Impact printers are the most common printer used for producing text from word processing systems. They also produce the best quality print, other than by phototypesetting. As they rely on a considerable electromechanical element, however, impact printers are generally slower, noisier and less reliable than non-impact techniques.

Impact printers use either *serial* or line printing techniques. The serial method prints one character at a time and generally produce better quality but slower speeds than line printers, which print a line at a time. Serial printers operate at speeds from as little as 10 or 15 characters per second (chps) to many hundred. Line printers can deliver up to 3,000 lines per minute, with over 100 characters in each line.

The IBM Selectric II golfball mechanism, which prints to about 15chps, is one of the oldest forms of serial print mechanisms. Two other techniques are now in widespread use, the *daisy wheel* and *dot matrix* printers. With a daisy wheel, up to 128 type character can be placed at the ends of spokes ('petals') on a rotating hub. The dot matrix mechanism consists of tiny printheads (sometimes called *needles*) which form a matrix of points to produce the required character. Typically the characters are formed with from between 5×7 matrix points to 9×14; the formation of each character is improved by using more needles.

Complete character printers like the daisy wheel tend to be slower but of higher quality than dot matrix systems. Typical prices of impact printers range from a few hundred to a few thousand pounds. There are over 50 manufacturers offering daisy wheel printers. The major producers of daisy wheel printers in 1981 were Qume, Diablo and Ricoh and these were fitted to the majority of word processors.

Line printers use a variety of mechanisms to carry the type characters such as drums, belts, chains and trains. The quality of the print is usually not good enough for letters and other mail. Their main text processing applications are in shared resource word processing environments, for example to produce early drafts of a document. Line printer prices are in the upper bracket for impact printers.

The use of microprocessors inside printers can provide a degree of intelligence for printing operations. For example, some printers can produce text that is automatically lined up (justified) to the right hand margin and there are dot matrix printers which are able to go over the same line more than once to improve the print quality.

Non-impact printers

Three types of non-impact printers were in significant use in offices by 1981: *thermal, electrosensitive* and *ink jet*. A fourth technique, lasers, was becoming more popular.

Thermal printers use similar matrix character creation techniques to the dot matrix impact printer, except that the characters are created by the application of heat fields on special thermal paper. Paper costs vary according to the degree of contrast and quality required. These devices typically have a 5×7 character matrix, operate at speeds slightly below those of daisy wheel printers and are available at less than £1,000. In operation they are very quiet, which can be a major advantage in an office.

Electro-sensitive printers also use matrix techniques, this time by applying an electric field to special paper. Although the print quality is better and the speed faster than the thermal method, the strange texture and look of the paper can be a drawback, particularly as it is relatively expensive. They use typically 5×7 and 9×12 matrixes.

Ink jet printers do not need any special paper. With some ink dot printers, a continuous jet of ink is pumped, which breaks down into droplets that are deflected to form the required shapes when they pass through electrostatic plates. Impulse jet printers squirt ink at the paper only when needed. Higher speed ink jet printers fire up to 200 jets per inch at one spot on the paper, with the character formed by the pattern of on/off jets.

Ink jet techniques can print on glass, plastic and metal as well as paper. At higher speeds, however, the characters often smudge and the ink jets can frequently become clogged up.

Laser printers are controlled by microcomputers and act in a similar fashion to xerography photo copying techniques but their price/performance needs to improve considerably before they will come into widespread office use. Non-impact printers can generally produce only one copy at a time because there is no impact to make an impression on copies beneath the top one.

Summary of important printer capabilities

The following are some important printer capabilities that should be examined when selecting a word processing system:

- The printer should be able to act independently of the editing function so that one document can be printed while the next is being edited.

- A range of options of printer devices and speeds may be needed.

- Many typefaces may be required from the same device.

- Special paper handling requirements, such as printing on continuous stationery and accepting automatic single-sheet feed, should be specified.

- Special ribbons and other print media may be required.

- In the event of a breakdown or other interruption, printing should restart from the top of a page or document and, if available, output should be directed automatically to another printer.

- Proportional spacing for printer output may be a desirable option.

- When many documents are waiting in a queue to be printed, it may be important to be able to have a list of the items in the queue displayed on the screen and, if necessary, to remove any document from the queue without affecting the printing of the remaining items.

Reliability, interconnection and general environment

A word processor should be capable not just of carrying out the tasks allocated to it quickly and efficiently but should do so reliably and over a long period of time. It should, therefore, have the potential to be enhanced and interconnected to other systems as the longer-term strategic plans unfold.

The following is a summary of some of the key reliability, interconnection, hardware and environmental characteristics that should be considered:

Reliability

- *Mean* (average) time that can be expected between total systems failures.

- Mean time between failures of individual components (processor, storage, printers, communications, interface).

- Mean time to repair significant faults.

- *Graceful degradation* capability — can working be realistically continued when one or more components fail?

- *Disaster quotient* — how often has text been lost irretrievably and what quantities were lost?

- When there is a system breakdown (a system *crash*), how much information is likely to be lost, what facilities are provided to recover from the break and how quickly is live working likely to start again?

Hardware and environment specifications

- Dimensions weight of system modules. Will they all fit into a standard lift?

- Heat emission levels for all system elements.

- Cable diameters.

- Power supply requirements.

- Cooling and air conditioning requirements to avoid making life uncomfortable in the immediate vicinity, particularly during any extreme weather conditions.

- Noise levels, particularly of printers, and any requirement for noise suppression equipment.

- Air filtration.

- Anti-static provisions.

- Special cleaning for equipment and environment.

- Smoothed power to avoid variations which can create flicker on the screen and other equipment abberations.

- Standby power supplies if continuous operation is important.

Looking to the future

Screen-based word processors will continue to be the growing means of carrying out text processing activities in the office. They will erode the market for conventional typewriters and take over in those activities where editing of text is heaviest.

In other areas, word processors may be used in conjunction with conventional typewriters, for example with documents typed initially using special type fonts that can be recognised by OCR input equipment.

Generally, what is wanted from word processors is: MORE! More types of systems, more advanced editing software, more communicating word processor facilities, more links to telex, viewdata, voice input, bubble memories, phototypesetters.

△
Word processors are the cornerstones of electronic offices. Screen-based systems like the Nexos 2200 improve typing productivity and speed up preparation and modification of documents.

◁
The thin optical fibre lines around the shoulders of the British Telecom engineer can carry as much information, more reliably and efficiently than the traditional telephone cables shown on the drum in the background.

△
The Rediffusion R800/70 shows
the versatility of office
computers—screen/keyboard
VDU to communicate with
computer (processor and storage in
cabinets), TV for viewdata and pad
for direct handwriting input.

▷
Electronic typewriters, like this
Olivetti system, could be thought
of as sophisticated automatic
typewriters, providing limited basic
word processing capabilities at
relatively low cost.

△
A Philips videodisk which is used for entertainment and training purposes can also hold images of thousands of documents for automatic search and recall by computer.

▷
The Microwriter is a portable device developed in Britain which can record text using a special five key code plus a single line display.

◁
Facsimile transceivers can send and receive copies of documents via telecommunications links, acting as a 'telecopier'. This is the Nexos/Muirhead 6400 transceiver.

The Fujitsu 8-inch hard disk ('the Winchester') provides low-cost high capacity storage for office-based computers. One example of magnetic media used for electronic filing.
▽

7 Information moves on electronic wings

The importance of telecommunications

No office is an island. At some point, most information handled in an office will need to be communicated to an organisation, department or individual. Some of that communication will take place within that office but a great deal involves a link to offices and people elsewhere — in the same building, in another building within the same organisation (perhaps in another country) or with people, companies and organisations in the 'outside world'.

Telecommunications provides the electronic highways along which information can travel into and out of an office. Telecommunications speeds are so high that it could be regarded as giving to information communication the same boost that air travel gave to physical transportation.

The two most common means of long-distance information communications have been the telephone, which is a form of telecommunications, and the national and international mail services. In addition, of course, people travel to work in offices and to hold meetings in order to have an effective means of exchanging and discussing information. Developments in electronic digital information technology already discussed in this book have made it feasible for a variety of text, pictures and voice communications to be carried on the wings of telecommunications in analogue or digital form. Electronic communications also encompasses activities which have traditionally been the domain of mail, publishing and library services.

Telex, facsimile, communicating word processors and computerised message systems are examples of electronic mail. Viewdata and teletext are forms of electronic publishing and dissemination that overlap with TV as well as other publishing media. These forms of telecommunications are discussed later in this chapter.

They all have one common requirement, an adequate telecommunications network. The most well-known network is the public telephone system, which is sometimes called the Public Switched Telephone Network (PSTN). With this, connections are made between a caller and receiver by dialling an address code (the telephone number) which is switched by the exchange to the appropriate receiver. Addressing and switching are basic requirements for all networks, except where there are fixed direct links.

The PSTN has been built up as an analogue-based service but eventually digital transmission may be used. There are other public networks designed for specific types of service, such as the UK Packet Switched Service (PSS) which is a special way of transmitting computer information by parcelling data into packets. Instead of using an exchange to switch circuits to make a connection between caller and recipient, packet switching uses a coded address at the front of each little parcel of digital data as the means of leading it to its destination.

Some large organisations have set up their own private communications networks to handle data, text and other communications between different sites. These private networks can use lines leased from national telecommunications authorities, such as British Telecom, plus satellites and other links which can also interface with public networks like the PSTN.

Local networks can be built to carry information between offices within a single site. This typically consists of a high capacity 'wideband' cable of about 1 kilometre in length which can carry information electronically at many millions of bits per second. Many devices and systems (word processors, printers, computers, storage) can be plugged into the local network which is continually broadcasting information around the network. Facilities are also provided to enable devices and systems to 'listen' to information flowing in the network. When a 'listener' finds information addressed to it, the information is pulled out of the system and onto the relevant device or system. In the future, the scope, efficiency and reliability of available public and private networks will be of great significance to office information systems.

The electronic postman knocks

In 1980, British Telecom produced a simple advertisement with the message: Speak A Postcard. The concept of electronic mail could not be put more tersely. It can encompass any communication between two or more people in which the transmission is, at least in part, via electronic equipment.

In the advertisement, the illustration was of a telephone. Instead of writing and posting a postcard, make a telephone call, it said. Other alternatives to the postcard could be a telex message, facsimile transfer, communicating word processor link or an exchange of messages via computers. Except for telex, which has its own network, the other techniques also make use of the telephone network.

In the context of the electronic office, the term 'electronic mail' is applied to those techniques which can produce a physical copy of the message or document at the receiving end of the transmission. This section concentrates on such 'physical copy' techniques (telex, facsimile, word processing and computers) but the telephone should not be forgotten as an important form of electronic rail. Viewdata can also be used to send messages.

With more and more information being produced in a form suitable for electronic transmission, the potential for electronic mail will expand. The need can be broken down into three main categories: internal mail within a site; mail between sites which communicate regularly; and mail to and from the world at large.

Internal mail within a site is the information transfer over which a company has most control and scope. For the larger organisations in the CSA study, about threequarters of the mail analysed originated within the organisation, much of it on the same site. In order to handle the high volumes of internal electronic mail and other information communications activities, local networks of high speed and capacity are likely to be used in conjunction with any public telecommunications services. Communication networks for handling internal electronic mail within one site should be able to handle text, voice and pictures. The internal local network must be highly reliable and should make it easy to connect a variety of devices and systems to it without involving special expensive and complex interfacing. It should also have sufficient capacity to ensure that messages input to the system do not have to queue for more than a second or two before being transmitted.

An effective internal electronic mail system will need at least one workstation for each group of 10 to 15 office workers. In some cases far more will be needed because certain staff will secure personal access to a workstation for most of the time.

External mail to other sites in or outside the organisation will depend on the availability of suitable public services and the resources available to the organisation to invest in and develop large private networks on their own or in conjunction with others. Some countries have regulations which limit electronic mail between different organisations because the national Post and Telecommunications authorities have monopoly rights over mail.

The following is a more detailed look at particular electronic mail capabilities.

Telex is growing into teletex

Telex is a long established means of sending messages and the cost of hiring equipment is low. In its traditional form, telex is suitable for only short messages because the presentation of the message output is unattractive and the transmission speeds slow, traditionally only tens of characters a second compared to many thousands for other transmission techniques.

Telex does, however, have a strong advantage because it has an established world wide network with agreed standards of interconnection.

A new lease of life is being given to telex-type messages by the introduction of an enhanced, faster and more versatile system called teletex. (*Note*: This must not be confused with teletex*t*, which uses TV to transmit information.)

Teletex enables telex-type messages to be sent at up to about 300 characters a second with output on screens or higher quality printouts than traditional telex. The potential for telex and teletex will also be reinvigorated as it becomes more common to link word processors and telex systems. Teletex/telex will become part of an integrated electronic mail service, with messages capable of being stored, edited and retrieved like any other computerised information. Eventually, teletex is likely to supersede telex and there will be a considerable overlap between these text message services and communicating word processors and computer message systems.

Teletex involves only pure text messages. Viewdata (also called videotex) and facsimile enable diagrams, pictures and other images to be sent, as well as text.

Facsimile for sending documents

Facsimile (fax) enables a copy of a document to be produced at a point remote from the original via a telephone link, which is usually the public telephone network. By 1980 there were about 15,000 facsimile units in the UK. The image transmitted by fax could be a page of typescript or handwriting, a map, diagram, photograph or any other image which may be originated on paper.

The image to be transmitted is first converted into the appropriate analogue or digital signals used by the transmission network and must be recreated by the receiving device.

Although fax can transmit a document page in 30 seconds and the quality of the document transmission has been improving, the growth in fax during the 1970s was not as great as many had predicted. The CSA strategic studies found little immediate need in the organisations studied to increase their facsimile services.

The CCITT (international consultative committee for telephone and telegraph) has been attempting to establish fax standards, which would be a precondition for any manufacturer being granted a licence for fax equipment to be attached to the public network. These standards have, however, tended to follow manufacturers' equipment specifications rather than leading them, so there has been a high degree of incompatibility between manufacturers' systems. However, technical developments in fax might be stultified if there is a too rigid enforcement of standards too quickly.

Some countries, such as France and Japan, are attempting to bring the cost of facsimile down and to extend use by developing 'mass fax' systems. These will be installed in homes as well as offices and could be used by smaller businesses which need to send a few urgent letters a week. Fax developments in Japan have been pursued with particular vigour because of the need to transmit documents written in Japanese languages which are based on complex graphical characters.

There are also developments of public facsimile services which involve taking a document to a post office or facsimile centre which is then used to transmit the document to another similar centre or an organisation with fax receiving equipment.

The most common facsimile units which were used in the 1970s could copy a page in 4 minutes, or 6 minutes if a higher quality is required. The sharper clarity of the six minute system was needed if the received document has to be copied or retransmitted after corrections have been made.

The simplest form of fax breaks the page into a series of small squares and scans each square to determine whether it is light or dark; the 6-minute unit operates at 96×96 scans per inch and the 4-minute at 64×96 scans. For each page, from 4 to 8 million bits of information is sent for each tiny square. The greater the number of squares, the better the quality and reliability but the slower and more expensive is the unit. This information can be compressed by, for example, sending just the number of successive blank spaces rather than transmitting each one individually. Such compression techniques have made 30-second transmission per A4 page commercially feasible. Using satellite transmission, pages can be sent in only a few seconds because transmission speed as well as data compression can improve the performance of fax. The reliability of transmission using compressed techniques can be relatively low unless error checking procedures are used. The greater availability of fast digital transmission networks will improve the competitiveness of fax. A basic fax unit is operated by making a call to the receiver, usually over the public network. Switches on both transmitter and receiver must be set to indicate the line is being used to send image data not voices. When the transmission is completed, a buzzer will sound and the received copy can be checked. If there is no more to be transmitted, the call will be terminated and the switch reset to voice mode, otherwise the next page will be loaded and the transmitter and receiver switched back into action. More sophisticated

Table 7.1 Comparisons of some electronic mail capabilities

Capability	WP-to-WP*	Telex	Fax
Layout	Good	Limited	Good
Clarity	Good	Good	Variable
Speed	Fast	Medium	Usually slow
Graphics and form presentation	Limited	None	Good
Unattended operation	Unlikely	Yes	Rarely
Receiver availability	Limited	Widespread	Limited
Editing without re-keying	Good	Limited	Usually none
Overcoming 'bad' line connections	Adequate	Adequate	Good

*(WP = word processor)

equipment, however, eliminates the need for attendance by an operator at the receiving end of the link.

Provided the major standardisation and cost problems can be overcome, fax could become a natural companion to telex systems and/or a replacement for many photo copiers. The potential UK market could be ten or more times greater than the 15,000 units installed in 1980.

Facsimile has been successfully used for many tasks, such as transmitting vehicle loading instructions to warehouses, sending engineering drawings to remote sites (particularly valuable internationally where posts services are slow) and in some forms of internal electronic mail, as well as for newspaper publishing activities.

Fast fax is more economic to use than telex if there are large quantities of information to be transmitted, particularly if the information is being communicated internationally. One British organisation has recently installed a fast fax network which it expects will save (on transmission charges) an amount equivalent to its capital cost within a year. Further cost comparisons are provided in Table 7.1 above.

Communicating between word processors

Once information has been entered and stored in a word processor, it can be communicated for storage, display and printing at another word processor, provided suitable intercommunication capabilities are available.

For companies with their own private communications network using leased lines, this could offer great telecommunications savings. Lines used in the day for voice can be used overnight for mail produced on word processors with negligible operating costs. The Bank of America was one of the first organisations to use communicating word processors in this way and expects to make substantial savings with it.

All shared logic word processing workstations can communicate with a central store and exchange information, so that a document produced at one workstation can be displayed and, if necessary, edited at another. In practice, however, shared resource systems are not widely used for electronic mail.

The main constraints on the growth of communicating word processors are the availability of suitable interfacing systems to enable simple and cost effective interconnections to be made between different manufacturers' work processors.

Computer-based message systems

Communicating word processors are one example of computer message systems. Computing software, storage and processing are the basis of word processors as well as large computers. Computer communications systems have been available since the 1960s. Initially they consisted mainly of terminal devices linked to a large central computer which shared its time responding to requests from terminal users, as in an airline reservation system.

As hardware became cheaper and more sophisticated communications software and network control grew, a wide diversity of computer communications became available, such as a distributed processing systems with many small and large computers and terminals interlinked.

Most of these services have a specific data processing task to perform, such as taking orders, answering enquiries, receiving accounting information and so on.

The Arpa data communication network established by the US Department of Defense, was one of the first to provide 'electronic mail box' *store-and-forward* facilities. People at workstations connected to the network could input a message addressed to another workstation. This is stored until the recipient requests it to be forwarded to his or her workstation.

As with communicating word processors, one of the stumbling blocks in the way of increased use of computer message systems is equipment incompatibility. Eventually, there will be a merging of word processor and computer message systems into an integrated electronic mail and information service.

Comparing electronic mail capabilities

The cost/benefit evaluation for the use of alternative information communication methods (mail, private messenger, telex, fax, communicating word processors, computer message systems, physical travel to meetings) depends on a variety of factors. Capital investment in word processors and computers, for example, may be justified initially for applications other than electronic mail. The basic telecommunications network may also have been justified wholly or partly for, say, data communications over a private network. A local network might be justified in terms of the integration of a variety of workstations so the cost for each type will be difficult to determine.

Telex and fax, however, need to be justified in their own right as electronic mail devices before being perceived as part of an integrated service.

In weighing the costs and benefits of electronic mail, it is important to consider travel costs. Travel and office equipment budgets are usually dealt with independently but it should be borne in mind that much travel to meetings is based on the need to carry and exchange information that must be analysed by many people at the same time. Advanced 'computer conferencing' and other techniques could enable many different people in different locations to talk to each other with fax, say, being used to ensure that everyone has copies of the document. Some of these more advanced techniques are discussed under integrated office systems in the next chapter.

Table 7.1 page 101 provides a rough guide comparing the capabilities of telex, fax and word processor-to-word processor (WP-to-WP) communication. Over time, some of the capabilities for particular systems will grow but the Table provides a summary of their broad areas of suitability.

Electronic mail needs support

Telecommunications that allow the transmission of electronic mail will eventually undertake more than just the straight communication of, say, a telex message or fax document. These additional electronic mail support services will include, for example, store and forward electronic mail box capabilities.

Other support service might include 'stamping' each item to identify the date and time it was sent and ensuring that multi-page documents can be sent and be received. These types of extra facilities should be an intrinsic part of the basic network services provided publically or privately.

Information dissemination and electronic publishing

The traditional way of disseminating information to a large number of people has been through printed publishing, radio and TV. There are two general methods of gaining access to such information, either as a *passive* recipient of broadcast material or as an *active* searcher through libraries, directories, magazines and newspapers to seek particular information.

Broadcast TV is a passive medium because, other than selecting a channel or switching off, the viewer can only plug into the material that is being beamed out by the broadcasting station. The electronic text equivalent of this is called *teletext* (not to be confused with the 'super telex' teletex.) The BBC Ceefax and IBA Oracle teletext services, for example, can send hundreds of pages of information in the same way as they broadcast TV pictures. TV sets with suitable receivers can interpret the teletext information.

Viewdata is the electronic equivalent of 'active'

information dissemination. It 'publishes' information to many people who are likely to search through it as a person in a library might look for, say, a railway directory and then to look through a specific timetable within it.

A generic term of viewdata and teletext is *videotex*; viewdata is interactive videotex and teletext is broadcast videotex.

Teletext and viewdata offer considerable potential as aids to office information systems. The power and flexibility of viewdata, in particular, could be used to disseminate information within an organisation or between organisations involved in similar activities, as well as offering access to the vast amount of useful information in the public Prestel system.

In the UK, viewdata and teletext work to common standards: 24 lines of 40 characters per page, a standard character set and seven standard colours. The CCITT and other organisations are trying to ensure that there are international videotex standards.

Versatile viewdata

Viewdata is a flexible form of computer-based information retrieval system. It was originally differentiated from other information retrieval systems by the nature of its original objective, which was to provide a public information service to people in their homes. Viewdata has now developed into an efficient business information technique. Special viewdata characteristics include:

- A low-cost, familiar terminal with which to communicate (typically a modified TV set linked by an ordinary telephone line.).

- A simple, easy-to-understand method of searching for and recalling information from the computerised libraries (using, say, a small numeric keypad.).

- A relatively quick way of establishing a computerised library which is divided into 'pages' that can be searched by the user.

Viewdata systems, however, have extended beyond this basis. A variety of specially designed viewdata terminals are available more oriented to business needs, with business VDUs and full keyboards. Viewdata systems can also be linked into virtually any form of computerised information service provided suitable interfacing capabilities are available or are specifically developed. An important feature of viewdata is that people can send messages to the computer and to other users as well as requesting information and it therefore could be used as a form of electronic mail. A mail order company, for example, could put a brochure on Prestel. Clients could then order goods directly from the Prestel set; direct payment using credit facility messages could also be provided. This would eliminate the need to post the brochure and for the client to send in the order and payment by mail.

A private viewdata system within an organisation could be used for a similar purpose. For instance, a manufacturer could place a Prestel set with every dealer to provide up-to-date information and to accept orders. In an office, a viewdata set could be used by a manager or secretary to find out travel and accommodation details for a journey. At the same time, a private viewdata system could be used to hold management and business information, directories of company departments and sites, reports, parts lists, minutes of meetings, schedules, etc., which could be accessed from a viewdata set linked to this network. Such a system could be confined to one organisation or part of a *closed group* viewdata network made up of organisations carrying out related activities. An interesting viewdata application for the office is as an electronic notice board. Instead of getting notices typed and copied, then distributed to be stuck on notice boards, they could be input into a private viewdata system. Vacancy notices, social functions, company news, union notice correspondence, etc could be made available to staff very quickly through the viewdata set in the office, eliminating typing, copying and distribution costs.

Unlike some information retrieval systems, viewdata files are not updated instantaneously. This would have added too much to the complexity of the software controlling viewdata. Pages can be updated frequently during the day using input terminals directly linked to the system or, for larger volumes, by overnight bulk updating. As care must be taken in formatting a page for viewdata, it could take about 30 minutes for an operator to enter a page. Software is available, however, which can take information from a data processing file and prepare it for viewdata. Special editing and formatting terminals are also on the market for viewdata information providers.

An information provider can choose to make the information publically available or limited to a predetermined 'closed user group'.

The information in Prestel is organised in what is called a *tree structure*. A person searches through the various branches by means of index pages which offer the user a 'menu' of choices; the user keys in the number associated with the relevant menu item, which then displays the appropriate index or 'end' page in the next branch level of the tree.

There are hundreds of thousands of pages on Prestel from many information providers, ranging from travel details and stock market information to career opportunities and educational exercises. Many other countries have or will be introducing public viewdata services and there are many private viewdata systems in operation; Prestel in the UK was the first public service.

The familiarity and relatively low cost of a TV set as a computer terminal and simplicity of use has also led many suppliers of office and data processing systems to offer viewdata as an intrinsic part of office and data processing systems. In addition to the price of the adaptor for the TV set or for a purpose-built terminal, Prestel costs include the telephone call charge (local call plus a special extra cost for Prestel) and a per frame charge that can be made by an information provider. For a private viewdata service, there will be the investment needed to provide the network but there will probably be no call charge and no per-frame access cost.

Tune into teletext

Teletext was also developed first in the UK. The BBC Ceefax and IBA Oracle services broadcast information by using two spare lines of the normal TV transmission not used to transmit pictures. With a suitable adaptor in the TV and a keypad, a teletext user can select a page to be displayed.

A teletext user cannot transmit any information back to the teletext computer, other than the page identifications. Compared with viewdata, there is only a limited number of pages available — a few hundred in contrast to the hundreds of thousand on Prestel. More pages could be shown if a complete broadcast channel was given over to teletext exclusively.

An important feature of teletext, which gives it an advantage over viewdata for highly volatile information, is that its information can be changed quickly. The information on a user's screen will be updated immediately if that page is being displayed or the latest update will be given the next time that page is recalled. This means it can be used for information such as news items, sports results, etc. A private teletext system could be used by an organisation, such as an airline, which needs to provide its staff, agents and outsiders with information that might frequently be changing.

The information output by teletext is controlled solely from the source of the broadcast signals but it could encompass information from a variety of sources which are edited and electronically published, in a similar way to a newspaper. TV companies have used teletext to provide subtitles for the deaf to certain programmes. Teletext could be broadcast through the air waves or cable TV.

Pages are transmitted by Ceefax and Oracle at a rate of about four per second in a continuous stream that runs through all the pages and then starts the cycle at the beginning again. When a user keys in a page number, the set listens in to the transmission until that page is being sent and then displays it almost instantaneously. On average, a user waits for half the total time it takes to transmit all the pages, which is why the number of pages is limited; otherwise users might have to wait many minutes. For a private teletext system which can use all the lines on the screen for transmission, the speed of transmitting a batch of pages is vastly improved, so the capacity can be increased.

Table 7.2   Comparison of viewdata and teletext facilities

Capability	Viewdata	Teletext
Time awaiting page transmission	Negligible unless system is heavily overloaded	Ceefax/Oracle average about 12 secs, depending on number of pages broadcast; private systems depend on design chosen.
Time to build up a page of information	Average 3-4 secs, maximum 8 secs	Negligible
Number of frames available	Large (could be hundreds of thousand)	Small (a few hundred) on Ceefax/Oracle but could be larger on private systems
Use of each receiver	Occasional	Frequent (over 25% of the time)
Frequency of page updating	Rare	Often
Interactive facilities	Yes	No

Comparing viewdata and teletext

Although viewdata and teletext are both electronic means of mass information dissemination, they are likely to appeal to different types of use. Viewdata is a fully fledged computer information service providing two-way interactive dialogues, electronic mail, access to data processing files. In the future, viewdata systems are likely to extend into other information areas, such as being linked to telex services and to computerised information library networks, using purpose built terminals incorporating computing intelligence instead of TV sets. Teletext, however, has a relatively limited role pumping out information that is regularly updated.

Table 7.2 summarises some of the differences between

viewdata and teletext. One point worth highlighting is that the access time to teletext pages does not vary according to the number of sets tuned in whereas viewdata response times can lengthen significantly if the system is overloaded with more users linked than its computing capabilities can handle.

Both viewdata and teletext are well developed and tested public services. Large companies could also consider private systems, for example to handle embryo electronic mail, or electronic notice board services. A start on a private viewdata system, for example, could be made modestly by hiring a Prestel set and some pages from an existing Information Provider to gain a feel of the system's potential.

All Prestel sets can also receive Oracle and Ceefax. It is, however, easier to set up a private viewdata service at remote sites than a private teletext service.

The winged electronic messenger

In 1980, British Telecom initiated a project called Operation Mercury. It aims to overcome the major stumbling block to the growth of electronic information services — the electronic Tower of Babel in which incompatible systems have great difficulty talking to each other.

Other similar projects are being worked on by public and private bodies and organisations throughout the world. Eventually, the chains of non-standardisation shackling the electronic winged messenger will be released and electronic information services will take flight.

8 File it! Find it!

Rethinking office system routines

Word processing and electronic mail are the two major focal points for innovation in office information systems. Running through all office work, however, are the familiar cries of File It! and Find It! Storing information and then finding it again is a common feature of all offices.

The availability of suitable digital information storage media is a vital factor in word processing, electronic mail and other computer-based information handling. Replacing the traditional function of the office filing cabinet, however, involves more than just the introduction of new storage media. It involves a new, systematic approach to the whole question of information storage and retrieval which needs to draw on the experience of specialist 'librarian' and computing skills.

In describing the general systems approach to electronic filing and library systems, this chapter provides an insight into the new types of systems thinking needed to bridge from conventional office routines to the Office of the Future.

Information storage and retrieval

A filing cabinet usually does not contain a random jumble of letters, memos, and files. If it did, the office would be extremely inefficient and work would grind to a halt each time a document had to be found. Instead, the files are organised and structured in a way that will assist filing and finding relevant information.

The manual filing and information retrieval system is not the only one. There is also the natural process of human memory and interaction. This natural form of information retrieval is also the most fruitful and convenient. No special techniques or procedures need to be learnt. The information stored in the mind has been gathered naturally and not specifically for the purpose of information retrieval. Most importantly, the human brain is adept at forming associations and developing leads and opportunities. The aim of any manual or automated office information systems should be to increase the effectiveness of 'natural' information retrieval through efficient means of communication and of accessing information when required.

Manual systems become necessary once information is committed to some tangible form, such as paper. As the volume of paper information grows, the first reaction is usually to classify and divide this base or corpus of information into manageable units and then to organise it into some defined sequence. The second stage is to develop indexes, at the minimum in the forms of labels on filing cabinet and manilla folders. If there is a lot of information in the files, special directories may be produced and it might eventually evolve into a fully blown library or large office filing system.

The basic data, such as that on pieces of paper, are called *primary* information. Indexes, colour-coded filing schemes, library card catalogues and other information needed to gain access to the primary source are called *secondary* information. Secondary information could be implicitly integrated with the primary information, say the index in a book, or else it may be completely independent, as with a library catalogue. Each item of primary information should, however, have some identifying key that could be used to find it, such as a letter reference number, customer name for a customer file index, or book title.

Independent secondary information tends to be generated once the volume of primary information exceeds the capabilities of the human memory. The creation of high-quality secondary information requires a thorough understanding of the primary material and knowledge of techniques for structuring it efficiently.

Office systems vary widely from organisation to organisation, even between offices within a department. Nevertheless some typical characteristics can be identified. The type of material in office filing systems encompass letters, memos, invoices, orders, cheques, payslips, reports, production schedules, engineering and technical drawings, publicity material, reference books and articles, personnel records, newspapers, magazines. It could take the form of:

      typescript
      manuscript
      printed matter
      computer printout
      forms and record cards
      photo copies
      drawings
      photographs
      microfilm and microfiche (microfilm 'cards')

The sources of information may be mainly from within the organisation but also from outside. The volume of information is likely to be large in terms of the number of discrete, separate items. Some items may themselves be lengthy, such as books and reports, but most will be short.

The active lifespan of information varies from a few hours for a brief message, to a few weeks for monthly reports, and decades for contracts and specifications. The prime method of organising this information base is to group items into files. Files are stored in filing cabinets, cupboards, bookshelves, desk drawers. Some files are maintained by individuals for their own use or for general office, section, division, department of corporate access. Thus there is likely to be a hierarchy of many filing systems dispersed throughout the organisation.

File usage includes updating and consulting information in them. Sometimes information is taken out for consultation (and may subsequently not be put back or put back in the wrong place.) For ease of file updating, information in each file is often kept in chronological order so that recent material can be added before older information. A single person or group of people may be responsible for updating files and controlling access to them. On the other hand, some files may be accessed by many people but without tight control in co-ordinating the updating and maintenance of the files.

Secondary information, such as classification and indexing of material, tends to be superficial. Most files are given a title which is written onto the file itself. Beyond that, an index may be maintained for a particular group of files. Some organisations impose filing schemes but these tend to be broad guidelines with considerable scope for local interpretation and variation. Indexing is most often controlled by secretaries and other non-specialist staff. Subject classifications normally have considerable variations.

Information is duplicated in the filing system by having the same item copied and inserted into different files. It is less common to have the same file held in many places.

The privacy and confidentiality of information is generally managed by physically restricting access to filing cabinets and locations through locks and keys.

This type of office filing system is so familiar that its shortcomings are rarely considered critical. We have grown to accept the many faults inherent in such systems, although more effective methods are available.

Problems with traditional office filing

In terms of information retrieval efficiency, there are three main faults with manual filing systems.

Firstly there is a lack of detailed, accurate corporate awareness of what information is available in its own filing systems. Where file indexes are available, the entries often include a vague word like 'General'. The indexes are often meaningful only to those who initiated them but are usually too cryptic for widespread use. This lack of corporate awareness leads to wasteful duplication of information and to managers and staff working with incomplete and/or obsolete information.

Secondly, the chronological sequence which eases updating is not always the most efficient for information retrieval purposes. A required item might be deep down in a file, for example, and it could be laborious to extract it/replace the succeeding items. The whole file could be taken if the retrieval task seems too time consuming but this creates problems when files are shared.

Thirdly, the absence of the individual(s) responsible for a particular filing system can effectively inhibit use of the system as a whole because nobody else really understands how to find their way through it. Such absence may be brief (lunch, coffee) or prolonged (sickness, resignation). The effect is shown by the time taken for new recruits to become familiar with sources of corporate information — it could take many months. Difficulties with accessing information also wastes time of executives waiting for information or trying to retrieve it themselves. Information can be lost through incorrect filing or because the prospect of returning something to the file leads to documents being left on desks rather than being put back where they belong.

There is clearly a great deal of improvement that could be introduced to office information retrieval. Computer-based services have shown that automatic systems can provide great benefits when the information storage can be optimised for retrieval activities as well as being able to handle updating operations. (See Information Retrieval Costs and Benefits, page 119).

Electronic filing and information retrieval systems were rarely in use in offices in 1980. There are a number of practical reasons why automated systems have been slow to move into office environments but there has also been a lack of understanding of computerised information retrieval.

Obstacles to automated office filing

In an office environment, any proposed automated system is competing with established processes incorporating filing cabinets, archives, librarians, filing clerks, secretaries and originators of information. Despite the limitations in being able to respond quickly, accurately and completely to a wide range of enquiries, manual systems can provide a flexible and reasonably priced means of accommodating material in many forms — text, graphics, drawings, manuscripts. The task of converting information from existing manual files to a form which can be used in electronic systems could be formidable.

Early computer and word processing systems did not encourage application to office filing systems. The cost of keeping information on magnetic media was higher than paper or film storage. Early word processors used magnetic cards, a low capacity storage technique which presented even more filing problems than paper. Few staff had screens or printers with which to look up information stored on magnetic media. And there was a general concern about the unreliability of computers and the lack of security and privacy.

In addition, the computerised information library and retrieval systems which had been developed during the 1970s were inappropriate to office work as they depended on having large computers with complex software to organise and manage the information base (*database*.) With an internal system, there is little opportunity for cost sharing as could happen with the provision of computerised library systems available to many organisations, clients and individuals outside the owning organisation.

The early automated information retrieval systems generally relied on skilled operation and development by experienced librarians and information scientists. These systems also used a more generalised notion of database organisation which did not feature the traditional office concept of discrete files.

However, many of the barriers to using electronic office filing systems have been progressively overcome.

The growing use of word processors with storage techniques other than magnetic cards means that more information will be generated in a form suitable for electronic filing and retrieval. Optical character recognition devices are becoming cheaper and more flexible so that it will be easier and quicker to transfer paper documentation to electronic form, without the need to re-key the information. The cost of magnetic storage is continually dropping to become a cost-effective competitior to paper.

The lack of awareness of automated information retrieval possibilities is being overcome through systems like Prestel and the provision of basic information search capabilities on word processors. A well-designed word processor can also be used to prepare information for computerised filing by automatically eliciting relevant secondary information for use in the indexing system, such as author, subject, date for documents.

There is still some way to go before systems are available which can deal automatically with manuscripts, colour, drawings, within the same automated system at an acceptable cost, systems reliability and performance. The complex systems and software developed for large computer information retrieval systems also need to be honed down and simplified for a better orientation to office work.

The first steps to electronic filing and office systems are therefore likely to be small. Eventually when, for example, videodisks, direct handwriting input and other techniques are fully developed, there could be more ambitious integrated systems.

Automated information retrieval systems

Early computer-based information retrieval systems used the computer's ability to sort and print information to produce library catalogues and indexes more effectively than was possible with manual systems. The main growth of large scale automated information retrieval projects came with scientific and technical library services.

There are many commercial and public organisations which offer access to computerised libraries (databases) over direct telecommunications links. The Euronet Diane service, for example, provides access to many databases over a public European network and the Lockheed Dialog service based in the US contains over 20 million items of secondary information (such as document abstracts) covering many fields of activity. There are also specific information retrieval software products available for a wide range of computer systems.

Computerised databases may consist of the full primary information text or secondary information made up of relatively simple facts like document author and title, carefully defined key words in the text, or more complex *abstracts* (summaries) of the document contents.

Primary information is input to the database either as the byproduct of another activity, such as typing a document on a word processor, or entered as a specific data entry task in its own right. The secondary information is likely to require knowledge of the primary information and a skilled understanding of, for example, how to produce complete but succinct abstracts. It can be input using any appropriate data entry method.

The heart of the automated information retrieval process is the search/enquiry procedure. The search criteria could consist of the specification of a variety of factors linked by 'logical operators' such as AND, OR and NOT. For example, information may be sought about documents which concern

Customer A *AND* Product B *OR* Product C but *NOT* Product D.

The first stage of the enquiry could be to retrieve secondary information on all items that fulfill the search criteria and then to browse through the text of these selected items. The secondary information can be organised by information retrieval specialists in ways which cut storage space and improve the search efficiency.

If the database holds only secondary information, there may be a link to manual ordering procedures for requesting the primary copy. If primary information is stored, it can be presented directly on a screen or printer.

Database management

It must be emphasised that there are forms of computerised enquiry services other than this type of electronic filing or library system. There is a wide variety of database management software available which enables computerised information to be stored in a structured way that allows different items of information to be interlinked as required.

These database management systems were originally developed for large computers to handle complex information sources. Instead of regarding the information as being divided into files with the same item of information frequently duplicated in different files, the principle of database management systems is to store an item of information once only and then to use the search mechanisms to link up different items into a particular file. Details of an employees salary, for example, might be wanted for a personnel 'file' or a payroll 'file' depending on the context in which it is being retrieved or processed.

One of the main areas of development in database systems has been creating enquiry languages and facilities which are natural and easy to use by people without specialist training. Viewdata is an example of an extremely simple technique of information retrieval and the whole system was honed to make it easy to use. Systems are also available which can run on smaller computers and word processors to provide relatively limited information retrieval facilities compared to the comprehensive database management systems, but which could be of great benefit in offices.

Information retrieval costs and benefits

It is easier to provide a summary of the qualitative benefits of automated information retrieval systems than detailed quantitative justifications. System prices are continually changing, as are other factors like telecommunications and staff costs and the value of better quality information to improve management decision-making and operational effectiveness.

In contrasting the effectiveness of manual paper-based systems with a hypothetical automated system, the essential advantages offered by the automated system are:

- *Time*: Less time is required for the automated system to search a large volume of information exhaustively.

- *Effort*: The effort in searching for information is expended by the computer rather than people; searches can be undertaken that would not be practical using manual systems.

- *Completeness*: Given a correctly defined set of search criteria, the automated system can guarantee a complete search through available material; no items will be missed or overlooked.

- *Accuracy*: The automated system will help construct an accurate specification of search criteria and allow requirements to be altered or modified in the light of the response; searches can be retried repeatedly until the desired response is forthcoming.

- *Efficiency*: There will be a reduction in the amount of duplicated information

- *Coordination*: Corporate management can be more aware of available information and so improve information management coordination. This avoids clashes that can occur when more than one person wants access to the same file.

- *Staff independence*: File retrieval will not depend on the availability of particular staff responsible for the files.

- *Enquiry range*: The automated system should impose fewer constraints on the range of enquiries that can be handled by the system.

- *Bonus leads*: The automated system may incorporate features specifically intended to highlight associated information that could lead to related and important information not originally considered by the person making the search.

- *Refined search*: The ability to construct complex search criteria cuts down the amount of superfluous information returned to an enquiry.

The following are some general guidelines of particular cost aspects of automated systems:

Storage

There are three major storage costs: for the media (paper, magnetic tape), equipment (filing cabinet, disk unit) and floor space. A study which compared a standard four drawer filing cabinet (capacity 20,000 pages) with an 80 million character storage module (40,000 pages) and 300 million character disk (150,000) produced the following results, assuming average UK office space costs and other prices at 1981 values:

Storage	*Practical range of cost per page*
Filing cabinet	1p-2p
80 million character disk	6p-16p
300 million character disk	3p-6p

During the 1980s, it is confidently expected that computer storage prices on a cost per page measurement will become more competitive with paper.

Creating the information base

The cost of capturing information for input to the information base varies considerably according to whether it is obtained as a byproduct of another operation or specifically for information retrieval. In addition, when an automated system is first introduced, there is likely to be a significant initial conversion cost when transferring information from manual to computerised files.

The cost of data capture is negligible if, say, information is created as the result of typing a report on a word processor, where the costs would have been allocated to the word processor's typing activities. On completion, the text can be transferred automatically from the word processor store to the electronic filing system.

Where information is entered specifically into the filing database, costs could be a few pounds per A4 page. If the information is reduced to a database primarily of secondary information, such as indexes and abstracts, the total volume of stored data will be cut. Re-keying costs could therefore be reduced significantly, although there could be an additional cost if skilled staff is used. OCR equipment offers an intermediate technique between byproduct data capture and re-keying (see Chapter 9).

Information retrieval software

Costs of developing information retrieval for large computers are high. Typical costs for available software are between £500 to £2,000 a month (at 1981 prices). The development of new software tailor-made for the office environment, however, is likely to bring the costs down considerably. Only the largest organisations with a pool of skilled staff could contemplate developing their own special software.

External information retrieval services

The costs of accessing external information retrieval services vary considerably. Charges are made for the total time a user is connected to the service, the telecommunications usage and 'royalty' payments for accessing databases. For many organisations, Prestel is likely to be the first experience of an external information service but other database services will appeal to organisations involved in particular activities where access to certain scientific, technical, business or other information is of great importance to efficient operations.

Electronic filing moves into the office

Ultimately, integrated electronic filing will come to offices so that managers and staff can exploit the opportunities for:

- more powerful and comprehensive information retrieval
- relieving staff of filing chores
- reducing the multiple storage and copying of documents
- ensuring staff and management are working to the latest, up-to-date information

This will, like other aspects of electronic office systems, come in stages. New text information retrieval software will be able to accommodate traditional office filing structures and routines more closely than software designed for computer-based systems. In addition, new features will be introduced and the need for laboriously generated secondary information will decrease, perhaps being assigned more at the time the document is originated.

Gradually, information retrieval will merge with word processing, electronic mail, data processing, into a wholly integrated electronic office information system.

Part III
A window on tomorrow's office

9 A technological route map to the future

The electronic office takes shape

This book has been a journey. Starting from a general analysis of information activities in offices, it has shown, step by step, how new electronic information technology equipment and techniques can be used to improve the efficiency and effectiveness of office information systems. While following a central main road, the journey has branched off to visit particular aspects in more detail.

The point in the journey has now been reached to pause and take stock of the new ideas and concepts that have been introduced. A great deal of new technology has been introduced but the shape of office work still remains along traditional lines. This chapter summarises the ways in which electronic technology can be applied to the information procedures introduced in Chapter 1, as a guide to the likely technological road ahead. The way in which people, organisations and society at large may respond will be discussed in the next chapter.

To reiterate from Chapter 1, the main information activities in an office are information gathering and input; storage; processing; updating and retrieval; communication and distribution; and output. The technologies for each of these component activities are being gradual drawn together to create integrated electronic office information systems. Chapter 10 examines the integrated electronic office.

The following sections discuss the techniques for each activity. Equipment and techniques are examined in some detail only if they have not already been analysed earlier in the book.

Text entry: getting the information in

The most common way of entering (inputting) text into processing systems is via a keyboard similar to a typewriter. This is likely to stay the predominant method until at least well into the 1980s. The keyboard is usually associated with a display screen as a screen-based word processor or a VDU linked to a computer. The traditional QWERTY keyboard is the most popular for electronic text entry. Other forms of text entry include voice, handwriting, OCR, hand-held typewriters, like the Microwriter, and computer input from microfilm. In order to overcome management resistance to learning to use a keyboard, there is also likely to be a development of spacial executive terminals. Other forms of data processing input, such as telephone keypads, are also likely to find a use in text entry. Some text entry requires a combination of techniques, such as OCR which requires a document to be keyboarded before it can be read.

Talking to a computer

Equipment has been available on the market for many years which can recognise a limited vocabulary (perhaps as much as a few hundred words). Such systems, however, could not be used to dictate letters to a secretary in a free-flowing way as the equipment can only understand clearly enunciated words and phrases but not continuous natural speech. Voice input could complement a keyboard to provide commands like enter, file, stop, to perform input of numeric codes or to input short words and phrases to standard letters.

These early forms of voice recognition equipment usually have to be 'taught' the words that it is to understand. The less sophisticated equipment can recognise words spoken only by the person who taught it but more flexible systems automatically type out words spoken from a particular scientific vocabularly. The performance of prototype versions of the more flexible systems were inadequate for widespread commercial use in 1981.

It is feasible, however, that direct voice input may become a viable alternative to keyboarding during the 1980s.

A substantial benefit from voice recognition systems is that the person inputting text has both hands free for other tasks.

Developments of systems to allow direct voice input over the telephone could provide major efficiency improvements by eliminating the intermediary stage of writing down messages and re-keying them. A typical spoken vocabulably for an adult is about 5,000 words and voice recognition systems must approach this for continuous natural speech before this technique becomes a serious rival to keyboarding.

With integrated office systems, voice input can be used to send voice-memos or voicegrams. This does not involve any computer recognition of the meaning of the spoken words; it just encodes the speech° and stores it for replaying when required.

Understanding handwriting automatically

The CSA strategic studies discovered that a great deal of typed information originated in handwritten form. Direct handwriting input would therefore have a significant potential, particularly as part of executive terminals.

The early handwriting recognition systems, however, were oriented to clerical tasks. The Quest Micropad, for example, comprises a writing station and 40 character display. The information is written with an ordinary pen or pencil, usually onto a preprinted form and using block capitals only. A microprocessor is used to analyse the handwriting and translate it into an appropriate form to transmit directly to a computer. Responses and prompts from the computer can be shown on the display. A similar device, the Image Data Tablet, recognises handwriting as it is written, based on providing current through the pen and writing on a special surface.

A system has also been developed to verify signatures when written on an electronic pad.

Two examples of handwriting input were recommended in the CSA studies which exemplify distinct requirements. One is for a system to recognise handwritten documents so that they can be input, stored and retrieved without any processing. Another was for use at a 'point of sale' checkout in a retail store, in which case the aim is to encapture data in an encoded form for subsequent computer processing and analysis.

Handwriting input will be more widely applicable when

equipment understands lower case as well as upper case and also joined-up, natural handwriting not just discrete letters.

Optical character recognition

OCR devices are use to convert printed, typed or computer printed characters into digitally encoded electronic equivalents.

In early systems, the documents had to be typed or printed using special OCR A (American) or OCR B (European) typefaces. They were initially developed to avoid the data preparation bottleneck, especially for airline ticket accounting systems. The banks generally used MICR (magnetic ink character recognition) on cheques.

From a cost of about £150,000 per reader in the 1960s, OCR equipment costs have fallen, in some cases below £10,000, while the versatility of systems have greatly improved. The CSA strategic case studies found, however, that most available equipment was still too expensive or too restricted in terms of the range of typeface fonts that could be read to meet the requirements identified. One of the main needs for OCR is in situations involving filing and archiving activities which would also need some form of input of images as well as text.

Many OCR readers work by scanning the text and converting black and white variations into electrical signals which are compared to a pattern of known character dot matrix images. More advanced and flexible systems analyse each character into topological and geometric properties such as loops, concavities, line segments. This enables the machine to read a variety of type fonts intermixed within a document or book and to cope with substandard print containing fragmented, broken or joined characters.

Kurzweil (subsequently taken over by Xerox) introduced such a multi-font OCR reader in 1979. It was originally developed for reading books to the blind and converting the output into synthetic speech. The same technique could also be applied to converting printed material directly into electronic storage for subsequent text processing, information retrieval or to input incoming mail into an electronic filing cabinet. If an organisation cannot afford to own such a machine, access to one could be obtained via an OCR bureau machine run by another organisation.

Cheaper, more limited OCR equipment has been used by many organisations in conjunction with word processing. The main attractions are:

- *Relatively low cost input*: Ordinary electric typewriters can be used for initial input at a capital cost about one-tenth the price of a full word processing key station or VDU; such cost savings will obviously depend on the volume of work that can be handled by OCR and the cost of the OCR system.

- *Easier distribution of word processing*: Since an ordinary electric typewriter may be used for input, existing secretarial groupings can be used, which means less organisational disruption and staff training.

- *Better use of word processing equipment*: Although input typing from a word processor is usually up to 15% faster than a typewriter, the word processor comes into its own on revision, updating, and other text editing activities. An OCR reader can therefore free the word processor operator from less productive input typing.

- *Media conversion*: The OCR reader enables work prepared on earlier word processing machines using magnetic card or paper tape media to be input without re-keying.

Portable electronic typewriters

The Microwriter, developed in the UK, was the first example of a portable electronic typewriter for use in word processing. It has only six keys, one of them a control key. Each ordinary character is formed from pressing a combination of the microwriter's keys. Its memory can store up to eight pages of typescript, which can subsequently be printed on a letter-quality printer and/or transferred to normal audio cassette tape for additional storage.

The limited keyboard means that the Microwriter user must learn a special code of key combinations to produce normal alphabetic characters, numerals, punctuation and layout instructions. Such training, however, takes only a few hours.

During typing, the character input is displayed on a strip window display. As the device is shaped to fit in the hand, the compact unit is convenient for carrying around and could be used as an alternative to dictating equipment, producing information automatically in electronic storage form.

In early 1981, Sony announced the Typerecorder, a portable system with a full QWERTY keyboard incorporating a single-line display, memory and mini-cassette. Other portable electronic systems are likely to emerge.

Computer input from microfilm

The Department of Health and Social Security in Newcastle was the first UK user of CIM equipment. Forms are filled in by hand by clerks in local offices. They are microfilmed using high-density recording and read into a central computer for subsequent processing. The system works effectively but its high initial cost (about £½ million for the DHSS project) means that its use is limited to large organisations with extremely high volumes of information to be processed.

Telephone keypads

Direct data input to computer systems using small keyboards linked to a telephone line are in widespread use. For example, many motor manufacturers supply dealers with a keypad for entering orders automatically with the responses from the computer and the prompts to the dealer to input particular messages being given by automatic voice response systems. Such keypads usually comprise digits and a few control keys. They have been used primarily for data processing, although such systems could have applications for text processing.

Executive terminals

Many businessman, managers and professionals are reluctant to use conventional QWERTY-type alphanumeric keyboards. This is due to sociological concepts of the 'status' of keyboarding as well as a lack of training. The Microwriter is one attempt to find an input device acceptable to executives.

Another interesting possibility is for special workstations, called executive terminals. These consist of limited voice input (mainly brief commands), handwriting recognition, special function keys, which will appeal to managers. Better software facilities are also needed to appeal to executive-level activities which, together with executive terminals, could fulfil the belief of the CSA study that executive usage will be the area of prime cost benefit in the future.

Non-text input

There are many other forms of entering information to computer systems which do not have wide application as first line text input devices but could form part of integrated information systems that also incorporate text processing.

There are systems, for example, used in retail outlets to read codes preprinted on goods by special *light wands*. There are also *touch-sensitive* screens which respond directly to finger-tip pressure and light pens and cats and mice (see Chapter 6) are available for use with screen displays and graphics. Digitisers and image scanners can translate drawings, diagrams, photographs and other images into digital form. Sensing devices can record environmental conditions both for help in direct computer control, for example over temperature and air conditioning, as well as collecting information for subsequent analysis and presentation. The technology is moving rapidly towards being capable of handling a wide variety of information input. The main difficulties are doing so at cost-effective prices with adequate performance and reliability and then of integrating various techniques into a single system.

Future trends of text entry

Most text will continue to be entered into computer systems via keyboards until at least the mid-1980s, with electronic keyboards increasing their share of the market. The main innovative trend will be towards speech and handwriting recognition and the availability of more devices, many of them portable, which appeal more to managers and other 'authors' who generate material rather than being aimed at secretaries and typists. OCR will also become an increasingly

Table 9.1 Guidelines of cost of text entry devices

Device	Price
QWERTY Keyboard	Up to a few hundred pounds
Dvorak Keyboard	Similar but slightly more expensive
Maltron Keyboard	About £500
Voice recognition	*About £6,000
Handwriting recognition	*About £1,600 to £2,000 per unit
OCR	**About £10,000 to £80,000
Microwriter	*About £400
CIM	**£500,000

*Includes input device plus other capabilities, such as processing power, needed to produce information ready to be stored and processed or printed.
**Centralised systems capable of coping with input from many individuals rather than just one at a time.

important means of entry into electronic mail and other information systems.

Information storage

Much of the text which eventually passes through text processing systems is still stored on paper. Microelectronics has brought down the costs of the relatively low-capacity, high-speed storage needed for computers' main memory. But paper has remained a more cost-effective means for bulk storage compared with magnetic media, such as hard and floppy disks, magnetic tape and cassettes and magnetic cards.

Silicon chip memories for the short-term storage of programs, text and data will continue to increase in storage capacity per chip and fall in price during the 1980s at a fast pace.

An important storage innovation that will start making a wide commercial impact in the 1980s will be bubble memories. They provide a compact, high capacity and fast access means of storing information which offer an alternative to floppy and hard disks. 'Bubble chips' containing 1 million bits were developed in 1979 and chip capacity of many times that are also feasible. This provides in intermediate form of storage between main memory and large capacity disk and archival store. Floppy disks, however, still have advantages in terms of media handling flexibility, for example when transporting data. Both floppy and hard disks will also continue to provide significant improvements in cost per bit.

For archival electronic filing which involves the long term retention and archiving of text and documents for relatively infrequent access, laser and electro-optic techniques have been developed, such as the the videodisk. It is unlikely that these will be available in cost effective form until towards the mid-1980s and then only using techniques which produce a permanent record that cannot be altered. This means their main applications will be for electronic filing rather than processing tasks. Billions of bits of information can be stored on a single videodisk, equivalent to tens of thousands of documents.

Microfilm is often chosen as an archive medium and COM (computer output to microfilm) can be useful in many circumstances. Microfilm is an image storage form which does not readily lend itself to computer processing although CIM techniques are available. It is therefore seen as remaining a parallel and complementary alternative to other computer storage techniques, until electronic archival techniques are available at competitive prices and performance. The availability of cheaper memory that brings storage for electronic systems into competition with paper will provide a spur to more imaginative and integrated information systems.

In addition to physical storage developments, the growth of electronic filing, updating and retrieval will depend on the availability of software to support these activities, such as database management and library information retrieval systems discussed in Chapter 8. The software access mechanisms developed for devices like the videodisk play a crucial role in determining the performance and cost effectiveness of a device.

Processing information

The availability of low cost, compact computer processing power in the form of silicon chips has extended the range and variety of processing equipment for use in office tasks. Microprocessors have been incorporated into many systems used in the office, such as word processors, desk-top computers, microcomputers, small business systems and minicomputers in addition to the large mainframe which has been used for many years in central data processing systems.

There is a great deal of overlap between the various classifications of systems. In general, for example, a micro-computer is smaller, cheaper but of more limited capability than a minicomputer, which in turn is smaller, cheaper and of more limited capacity than a mainframe. On the other hand, depending on the software and other capabilities, some minicomputers are more powerful and expensive than some smaller mainframes, while larger microcomputers overlap with smaller minicomputers. Some word processing systems incorporate their own minicomputers while some small business systems incorporate word processing facilities.

The most significant trend in processing capabilities is that the cheapness and compactness of microprocessors has led to a general possibility of putting intelligence (processing power) in equipment sited where work exists, rather than moving the work to a central unit. Such equipment can be linked together in a distributed processing network.

Processing hardware itself is unlikely to be an inhibitor to the growth of text processing, given the rapid and continuing improvements in price/performance. Microprocessors have been and will be incorporated in many pieces of equipment without the user being aware of their presence. For example, microprocessors exist in printers provided with screen-based word processors and in the decoders which modify an ordinary TV to receive Prestel.

Although hardware processing capabilities are continually improving, it is only when suitable software has been provided that benefits to the user really occur. Software is the root of applications flexibility. It is the software editing capabilities, for example, which determine the suitability of particular systems.

A limiting factor in the growth of text processing and other computing systems is the hardware and software incompatibilities between systems from different suppliers and even between systems from the same supplier. Processing hardware can be used by suppliers as a means to 'lock-in' users to their systems, particularly where word processing is offered on data processing systems and vice versa. These incompatibilities are being overcome with the development of communications networks which automatically handle many different types of devices and with 'interface' devices or software systems which act as a 'translator' between equipment.

In addition to further improvements with semiconductor silicon chip technology, other hardware developments are likely to pack more processing power into a smaller space and at lower cost, such as the Josephson Junction technology developed by IBM.

Most users will become aware of hardware processing developments in terms of the new (mainly software) capabilities provided rather than by explicit visibility of the processor itself.

Information retrieval and database management

In the traditional office, the focal point of the office 'information base' is the filing cabinet. The information database in the electronic office will provide an automated way of updating, searching through and retrieving information in a more efficient, reliable, accurate and comprehensive way than manual systems.

Although these database management techniques depend on developments in hardware storage, processing, input, output and communications developments, the potential scope for electronic filing and related applications depends primarily on software facilities. The database needs to be organised by software in a way that makes updating and retrieval as efficient and flexible as possible and ensures that recovery can be quick and complete in case of a system failure.

The essence of computerised filing and database management is to enable the same information source to be used for a variety of tasks and to be searched in a variety of ways. Databases and electronic files will therefore play a crucial role in integrated office information systems.

Internal office systems will also have 'ports' or 'gateways' enabling them to be linked to external information retrieval services such as the Prestel viewdata system and public and private computerised libraries.

For most offices, computerised information retrieval and electronic filing systems will grow slowly as a natural extension of the growth in the use of word processors and other office computing facilities. A computerised information base is the natural byproduct of word processing; basic database searching and management software is available for some word processors.

Integrated text databases are likely to be available before other electronic filing methods are incorporated, such as manuscript, colour and drawings.

Communicating and distributing information

The most common method of distributing information which has been prepared on text processing systems is still on paper, although there are extensive public and private networks for electronic mail facilities such as telex, facsimile, telephone, interlinked computer terminals and communicating word processors.

One of the most common means of disseminating information has been to 'get a photo copy'. A great deal of office equipment costs relate to reprographics. Traditionally office reprographics have been divided into duplicating systems (such as using a stencil or off-set printing plate) or copying. Xerography enabled the same device to be used for a few or many copies and led to an explosion in the number of copies produced.

Improvements in copiers have introduced automatic paper feeding, collation of output, speeds improved from about 10 copies a minute in the 1960s to over 100 a minute by 1980 and the capabilities for reduction or enlargement of originals, colour copying and the overlaying of a standard form on the original. As a great deal of employee time is lost in walking to, queuing for, using and returning from the copier as well as time spent stapling, addressing and posting copies, many organisations have realised that actual copying speed is less important than facilities like automatic collation and assistance in page feeding for multi-page documents.

Technological advances are unlikely to make a significant improvement in user benefits for stand-alone copiers.

The main drawback of photo copiers and other paper communications that rely on the mail for dissemination is the time taken to get a document from one point to another. The speed of communications is the major benefit provided by electronic information systems.

At first, communications systems are likely to develop for specific types of transmission, such as telephone, telex, data or text processing before being integrated. This is both because organisations will approach electronic information systems in a gradual evolutionary fashion and because of the problems of incompatibilities between various equipment and systems.

With more and more information being prepared electronically there will be increasing needs for high capacity transmission systems such as fibre optic and video cables used for piped television. The national telecommunications infrastructure provided by British Telecom and other telecommunications authorities and organisations will be of vital importance in promoting electronic communications, particularly if facilities are provided and standards implemented to overcome incompatibilities.

Digital transmission and computer-based switching will increasingly dominate telecommunications systems. Digital transmission is faster, more reliable and efficient than analogue methods and is capable of handling all types of information in a single system using computer switching, such as British Telecom's System X exchange. It is also more efficient and reliable than previous techniques enabling a wide range of user facilities to be created by relatively simple software modifications, such as automatic call rerouting and 'conferencing' where many telephone subscribers can be interlinked for simultaneous conversations.

Private telecommunications network incorporating satellites, microwave and other links will also be of great significance for larger organisations. For offices, one of the most important developments is in *local networks*, based on techniques such as the Cambridge Ring and Xerox Ethernet, which provide high speed, high capacity cable links. Up to 100 or more workstations, computers, databases can be plugged into a local network.

An important form of information dissemination will be viewdata (interactive videotex), and broadcast teletext which together provide new techniques of electronic publishing as well as information communication. Links to high speed printers and intelligent copiers which can print and collate large documents will also enable electronic communications to speed the distribution of paper documents as well as pure electronic output in the form of displays, voices, etc.

Digital communications can also be used for 'telewriting' which transmits handwriting (without any processing in between) from one terminal to another. This could be used as part of an electronic 'blackboard' for video conferencing which transmits live pictures and sounds between two or more sites. The growth of particular communications equipment and services will depend on three main factors: availability of devices with the required capability at a cost effective price; communication networks that can make the required capability available to many people; overcoming incompatibility problems.

As technologies converge, say into the intelligent copier which could include fax transmission, OCR input, high speed printers, text storage and editing, the dividing line between different techniques will merge and fade.

Digital information will become like a universal glue used in the links that join and integrate various electronic information technology techniques.

Output and information presentation

The main output from text processing is likely to be through visual display screens and printers. Most of the required capabilities for these two output media were satisfied by equipment widely available in 1981. The two main outstanding needs were for increased graphics capabilities and higher speed quality printing. These needs are likely to be met by technologies already being developed.

With screens and printers, the likely future developments will be in the extension or improvement of existing capabilities, such as faster or quieter printers, and reduced costs. Screens are likely to get flatter and bigger, offering the prospect of large wall displays. There is also likely to be a growth in the use of split screens and other techniques of presenting and manipulating information to make the screen seem more like a desk top containing many different documents. The graphics used in most offices were initially relatively limited, such as form layouts and pie and bar charts. This has grown into a demand for more complex graph plotter and graphics print capabilities to copy images off the screen.

Intelligent copiers could become a popular means of printing and distributing information, at first in larger organisations. Another growing printed output method is likely to be the use of phototypesetters.

The link between word processors and typesetters will make a significant impact in more organisations than traditional printing and publishing operations. The phototypesetter could become an increasingly popular module in the automated office. As phototypesetters are themselves computer based, they can provide similar editing flexibility as word processors for manipulating text and are generally more versatile and efficient than metal type techniques. The main difference in editing terms between word processors and phototypesetters is in the software handling of the kind of page layouts typical of a newspaper or magazine rather than letter and report formats.

For users of word processors, a direct link to a photo-typesetter means that efficient output can be produced that takes advantage of the quality and flexibility of printed material and also cuts out the need to reset information that was previously produced by a typist then sent to the printer for rekeyboarding. There are two main technical problems — incompatibilities between word processor and photo-typesetter codes and commands and the increased operational complexity of phototypesetting. There could also be problems when print unions argue over which staff should make the keystrokes that produce printed output.

Eventually, phototypesetters could become one unit of an organisation's 'print room' that handles the internal production of a variety of printed output, such as word processor printers to large phototypesetters.

A halfway house between using paper and screens to present information is microfilm or microfiche. The advantages over other techniques are that the cost of a microfilm reader can be cheaper than a display screen and copies of microfilm are cheaper than copying paper.

Microfilm, however, is generally incapable of being altered or updated, although techniques have been developed to allow for limited re-use and extension of microfiche files. Access to COM services can be obtained through COM bureaux.

A technique that is likely to grow and challenge screens and paper in many circumstances is direct voice output. This can be done either through completely synthesised, computer controlled voices or by recording a person saying selected words and phrases and then using the computer to put them together to produce the required output.

Voice output has been incorporated in many consumer products, such as the Texas Instruments Speak and Spell educational game, calculators, cookers and washing machines. In 1980, IBM announced a 'speaking' automatic typewriter designed for use by blind typists which could 'speak' as well as print characters. Voice response systems are also used in communications systems, for example advising car dealers of the availability of parts in response to enquiries input over telephone keypads.

Devices, called codecs, are available which convert voice signals into digital form and back again. These could have wide application in telecommunications networks.

The range and versatility of output from computer based systems will eventually come closer to the scope of natural human methods of communication.

The jigsaw picture is complete

All the major elements in electronic office systems during the 1980s have now been described. There are bound to be important technologies that emerge unexpectedly but it is unlikely that anything radically different from what has been described in this book will emerge in a form that can make a widespread commercial impact within a decade.

Organisations must base their strategies on foreseeable developments not hypothetical ideas. Short and medium term plans, in particular, must be based on investing in equipment and systems techniques which should last for up to five years and more. A prototype experimental system may make an interesting item in Tomorrow's World on TV but a user must ask questions about price, proven reliability of the system, commercial viability of the supplier, and many more practical requirements, even if the potential evokes a 'Gee Whiz' gasp.

When all the pieces of the jigsaw fit together into an integrated system, the overall impact is likely to create a very different image from traditional concepts of an office. The next chapter concludes the journey by taking a glimpse at the picture formed when all the jigsaw pieces interlock.

10 The new office landscape

Integrated office systems

All electronic office techniques are evolving towards the integrated office system. This aims to provide full information and electronic processing assistance to help workers at all levels in offices. Each member of staff would have access to a workstation, probably with a screen and some processing power. The workstation will have communications links with other units to enable it to perform functions such as:

- *Sending* and *receiving* messages, memos (electronic mail)
- *Retrieving* information from electronic filing systems and centralised data processing systems
- *Inserting* information in data processing systems
- *Editing* text
- *Conducting* conferences using computer networks
- *Comparing* diaries stored in electronic databases to arrange meetings without disturbing an individual
- *Access* to outside electronic information services, such as Prestel

To achieve this requires an extensive communication system and a high degree of planned compatibility between systems. Workstations could be tailored to personal needs; they do not have to be identical within an organisation, provided standardisation problems have been ironed out. In fact, given suitable communications link, there is no intrinsic reason why the workstation should not be in a person's home rather than an office.

This opens up the opportunity for radical changes in the location and nature of office work. Offices took their traditional form because it was the best way, given available technology, to share information management resources and to ensure that there was efficient co-ordination and communication of information between people employed to do related tasks. An integrated electronic network eliminates many of the physical constraints because, for example, an electronic 'filing cabinet' can be accessed from a workstation at home rather than needing the person to be at the same location, as with manual filing cabinets.

This chapter examines the kinds of facilities that make up an integrated office system and then briefly discusses how developments in office technology might influence wider social and work patterns.

The introduction of an integrated office system offers the possibility of a profound improvement in the way organisations operate at all levels of staff. Initially most electronic office workstations were aimed at supporting the tasks of clerical workers, such as typists at word processors or clerks inputting accounting data at VDUs linked to a computer. This, however, tackles only a small proportion of the total office efficiency problem.

An integrated office system must also provide technological support to executives and professional and technical staff in order to release more time and provide access to better information sources. This should help to improve their creative activity and the quality of decision-making.

Office staff can be broadly classified into 'executives' and 'support' staff. In practice, the distinction is not clear cut and many staff perform intermediate roles. These classifications are, however, a useful way of analysing office information tasks. Executive staff cover job titles such as director, manager, administrator, technical, scientific and professional specialists. Their primary duties are to develop, record and communicate ideas and to formulate and communicate decisions. In performing these primary duties, however, studies have shown that up to 70% of an executive's time is taken up with secondary, peripheral activities such as locating information, talking on the telephone, filing, attending and organising meetings.

A reduction in the time spent on such peripheral activities would allow more time for executives to fulfill their primary duties. This is a main aim of an integrated office system.

Many of the peripheral activities in which executive staff are involved are performed by support staff such as clerks, secretaries, typists, messengers. This type of indirect working means that the executive staff must spend time explaining requirements to support staff. This introduces the possibility of misunderstandings and reduces the opportunity for more positive interaction between all levels of staff.

Electronic office technology was first applied to support functions because the nature of these tasks was more structured and well defined rather than because they would necessarily yield the greatest cost savings. Executive staff usually form a higher percentage of staff costs than support staff.

Table 10.1 below illustrates the characteristics of each work classification.

Table 10.1 Characteristics of executive and support work

Executive	*Support*
Complex primary duties: creative decision-making	Relatively straightforward primary duties: process-oriented predictable
High cost	Lower cost
Generally unstructured activities	Well-defined, structured activities
Multiple concurrent tasks	Restricted, serial tasks
Many information sources	Fewer information sources
Communications flexible and vital	Communications less flexible/ vital
Great mobility	Less mobility
Contact with many work communities	Constrained communities
Majority of time spent on peripheral activities	Majority of time spent on primary duties/activities
Traditionally, little technological support	Traditionally, main target of technological support

An information system to support executive functions must provide a more subtle and complex range of integrated facilities. Senior staff have a wider span of control, internal and external contacts and areas of interest than most support staff. The system must therefore cater for a greater degree of ad hoc interaction and unstructured communications. This means that the telecommunications networks should be capable of handling a variety of information types (text, data, voice, image). In addition, workstations should appeal to executive level staff who might wish to carry out many functions from the same unit. This has led to the development of the *multi-function workstation* concept, which is also known as the *integrated workstation* or, in more specific forms, as an *executive terminal.*

In addition to providing a variety of input, output and storage media, the multi-function workstation should be supported by software which makes the integrated office system behave as closely as possible to traditional office procedures, as well as offering new facilities.

An integrated workstation, for example, might be based in a largish desk. On the top would be a screen, telephone, computer voice and handwriting recognition devices, a keyboard (possibly a modified version of the full QWERTY with special executive function keys), a light pen or other device to communicate graphic images directly to a computer, computer-controlled voice output and a space for inputting documents for facsimile transmission. Various electronic filing storage media could be held in the drawers and other storage units beneath the desk top. In order to make the workstation operate in a similar fashion to a traditional office desk, the screen could be split into a number of 'windows' to display many documents at the same time.

In the morning, the executive might 'open' his or her electronic mail box by speaking or keying-in a password. The mail box electronic storage will contain messages and documents transmitted electronically from other organisations, departments and individuals with a link to the office information system. The organisation's own 'electronic mail room' will have opened all the physical mail and sent copies of documents addressed to that executive via fax link to his or her electronic mail box storage.

When going through the electronic mail, the executive could use an 'electronic waste-basket' to remove any unnecessary documents; the electronic mail waste box will be deleted periodically by the system but if a document has been mistakenly 'thrown away' by the executive, there is a chance that it could be retrieved. Any electronic mail that requires urgent attention could be assigned a high priority which would automatically send it to be stored in an 'electronic In/Out' tray and will be near the top of the 'pile' when the executive takes action to work through the electronic In tray.

A special facility could be a *voicegram* or *voice memo*. If the executive wishes to append a note, an enquiry, a request for action etc, to an electronic mail memo, instead of scrawling a note in the margin of a document, he or she can send a voice message. While the memo is being displayed on a screen, for example, a key could be pressed on the work-station which would set an indicator at the place currently pointed at by the cursor. The executive would then speak the message to the voice recognition unit and transmit the memo plus voicegram electronically to the person who needs to respond to it. When the recipient finds this memo with a voicegram in his or her electronic mail, an indicator will be set to show that a voicegram is present. At the appropriate point in the text, the voicegram message will be spoken on the voice output unit.

Integrated office systems will initially be similar to traditional office procedures. Eventually, however, the power and flexibility of this new information medium should create a new attitude to organising office work at all levels, just as it took time for creative specialists to exploit the full potential of the cinema as their own experience and the capabilities of the technology extended it beyond traditional concepts of theatre and photography.

Origins and examples of integrated office systems

Like most aspects of the Office of the Future, the integrated information capability has its roots firmly in the past. Work on the concept started in the early 1960s at the Stanford Research Institute in California on a project called NLS, which was used by a community of 'knowledge workers' — managers, researchers, administrators, engineers and scientists.

In the early 1970s, the system became available over Arpanet, the data communications network run by the US

Department of Defense. The prototype NLS system suffered from the limitations of technology available at the time, such as relying on centralised computer processing power and relatively limited storage, telecommunications and text handling capabilities compared with the cheaper, more powerful functions brought about by microelectronics and other advances. Despite the constraints imposed by the available technology, the concepts pioneered by NLS still remain valid, particularly the idea of a multi-function workstation.

The NLS workstation, controlled by a powerful computer and linked to a datacommunications network, provided text entry and revision, electronic mail, local storage of information, access to external information services and special executive facilities such as maintaining a diary and aids to help schedule meetings.

Knowledge of the NLS systems soon spread and fostered a climate of opinion in which it is possible to consider an equivalent system for the office environment. At the same time, the increasing attention being given to word processing facilities permitted similar conclusions to be reached from a different viewpoint.

A screen-based, stand alone word processor contains many of the elements of an executive work station: screen, local storage and processing capability plus the ability to communicate with other information systems. Additional functions, like voice and handwriting recognition, become available as the respective technologies advance.

There is a significant difference, however, between an executive work station and a typing workstation. The difference lies in the type of functions that need to be performed. Although the same hardware units may be common between the two types of workstation, different software is needed to support the different types of tasks carried out. Software is the critical technological factor in determining the speed of development and widespread use of commercially acceptable integrated office systems.

A pioneering integrated office system used in a live commercial environment was introduced in the late 1970s in the US by Citibank. Called Lexar, the system was developed to serve a small group of executives and their secretaries, with each executive and secretary sharing a minicomputer and having a screen each as workstations. The main objective was to allow each executive to control nine subordinates instead of seven. The early workstations consisted of specially designed video display units, six microprocessors, silicon chip main memory, disk storage and a daisy wheel printer. The manager and secretary can work independently at their own display terminals but share access to stored information and to a communications interface to external services.

The main early innovations in Lexar workstations related to the provision of a variety of services from the workstations, such as text entry and revision, file management and directories to files, access to data processing systems, electronic 'waste basket' and 'in tray' capabilities and code word 'signature' authorisation. Later developments included a calculator facility, split screen capability and mass storage information retrieval. The scale of resources for such developments is indicated by the fact that the total systems effort was estimated as up to 40 man-years.

The US company, Tymshare, Inc., started marketing a commercial version of NLS in 1978, called Augment. By 1980 there were about 500 Augment workstations linked over the Tymshare data network. Augment facilities encompass a variety of output terminals, including COM, text entry and revision, limited graphics potential, electronic mail and spelling verification.

Xerox Corporation was also a pioneer in developing an integrated office system called Alto at its Palo Alto Research Center in California. The system was primarily an internal research and development project aimed at investigating ways of creating easy-to-use workstations. Part of the development was to work closely with children to design new ways of interacting with the system. In 1979, the US Office of the President introduced an Alto system on an experimental basis. The communications network used by Alto is the Xerox Ethernet, which is a system also widely used as a local network for interlinking office workstations within an office block.

A British integrated office development is Scrapbook which was developed at the National Physical Laboratory, with Triad Computer Systems taking over the marketing rights in the late 1970s. The system is based on minicomputers and provides facilities such as text entry and revision, message routeing between workstations, file management and directories, an electronic 'in tray', broadcast of messages to many workstations and an interface to a typesetter.

Costs and future trends

There are three main aspects to integrated office systems: the workstations, communications network and software that provides support to many of the capabilities. The costs will vary a great deal depending on the equipment and systems that have been built up for other purposes.

Local networks like Ethernet, for example, could be justified purely on the basis of linking standard text processing equipment such as word processors and printers to provide an internal office information and electronic mail service. Much of the equipment might also have been obtained for other specific purposes, rather than as part of an integrated office system as such.

The Citibank requirement of 40 man-years of effort to develop Lexar indicates that it could be a major task to attempt to create a tailor made system. Obtaining systems from suppliers could be within the budgets of only larger organisations. Scrapbook software, for example, could cost from about £20,000 to £80,000. Access to Tymshare's Augment system could cost up to about £20 a day excluding telecommunications costs. (These are at 1980 prices.)

As the integrated information system is the ultimate objective for most electronic office strategies, there is likely to be an increasingly varied and cost effective range of equipment and services on the market. For example, the Delta system from the US Delphi Corporation, which is marketed in the UK and Europe by Nexos Office Systems, provides a network for integrating the processing and transmission of voice, data, text and images. Ethernet, the Cambridge Ring and other local network facilities can also handle all forms of digital information and similar capabilities are offered by many suppliers of electronic office and computing systems.

The capabilities of executive terminals and workstations will also improve as both users and suppliers become aware of the importance of providing more efficient and effective aids to management and professional staff. In one experiment by the University of North Carolina, it was found that the use of executive-oriented systems could improve both the quality of decision making and reduce the time taken to reach decisions by almost 20%, with a saving of about a third in the total man/hours effort needed.

Is the end of the office nigh?

As illustrated earlier in this chapter, information technology could place considerable information handling power at the finger-tips of office staff, wherever they are located. Extrapolating technological potential, it is possible to imagine the majority of office work being carried out from workstations based in the home, thereby eliminating much of the need for large office blocks.

This speculation is a kind of science fiction because it is based on an idealised creation of a world that has no past and present, only a future. If it were possible to start from scratch and build a society based primarily on technological capabilities, then it might be feasible to create a completely new pattern of social and work organisation. The same type of optimistic and idealised technological assumptions lie behind projections that a 'paperless' Office of the Future could come to pass virtually overnight.

As has been emphasised throughout this book, human, social and organisational factors, investment in conventional equipment, and habits and procedures built up over generations will temper the cutting, leading edge of technology. The technology itself has many limitations, particular in terms of integrating and interlinking a variety of functions and equipment. There are also many problems and time delays in translating the potential of an experimental prototype into cost-effective, reliable and well-supported and maintained systems.

To have people working at home rather than in an office would involve radical and far reaching changes in personal, family and social behaviour. In other activities, information technology is also opening up new opportunities for organising educational, business, shopping and other activities over telecommunications links.

There are so many interacting factors which affect social, economic and political developments that to regard technological innovation as the prime force in determining the future would be foolish.

If (and there is a significant if about many technological developments), Information Technology fulfills its potential, the result will be significant changes in the nature of the working environment, in job design, and in the patterns of employment and productivity.

At all times, however, it must be remembered that technology should be designed to fit human aptitudes and needs, not vice versa. An office is more than just a place of work and information management. It is also an environment which flourishes on personal and social interactions between people. Electronic office systems could lead to negative human responses if the informal gregáriousness and human contact is automated out of the system. It would also probably lead to the creation of inefficient systems.

In many offices, for example, an informal information network exists which is often the springboard of creativity and job interest and satisfaction. A meeting over coffee or wandering into a room to have a chat with colleagues often leads to information exchanges which spark creative ideas. The essence of creativity is the chance association of new ideas and care must be taken to design information systems to match informal, ad hoc networks as well as following formalised information systems analysis.

To say that technology should meet the needs of people is easier said than done. It involves much more than a silicon chip or new hardware capabilities. It requires painstaking discussion and analysis of needs, systems trial and error, gradual gaining of experience and great investment in software and systems, in human skill and ingenuity.

From the clear evidence of technological trends, significant economic factors like higher oil prices, and the general sociological desire to live and work in smaller communities, it is reasonable to suggest that offices in the future may be less centralised. Local community office centres equipped with electronic office information capabilities are likely to be developed for use by many organisations.

The aim of this book, however, has been to give a realistic, practical picture of office developments during the 1980s and not to enter the sphere of futuristic speculation.

The end of the office is not nigh, despite the revolution in office technology which was set alight by the microchip. The future lies in the hand of people who work in and run offices. The power and strength of electronic information technology is that its software-centred heart can be moulded to meet and adapt to peoples' changing needs.

It is up to people to use that power to benefit individuals, business and society at large. And that is best done by understanding the nature of our new technological office workmate. The future can then take care of itself.

Appendix 1
Glossary—finding your way in the jargon jungle

Alphanumeric	Descriptive of a keyboard containing alphabetic (A to Z), numeric (0 to 9) and, usually, punctuation marks and control (eg shift) keys.
Archival storage	Files stored primarily for historical reasons; characterised by being of high volume but infrequently accessed. See **COM**.
Arithmetic mode	See **Calculator mode**.
ASCII	One code which uses a *byte* to represent all digits and alphabetic characters; abbreviation of American Standard Code for Information Interchange. See **BCD** and **EBCDIC**.
Automatic typewriter	A device which looks like a typewriter but incorporates magnetic storage such as card, tape or diskette but with limited (if any) *editing* capabilities and no electronic display.
AZERTY	Keyboard layout used for non-English European languages; first six alphabetic characters are A Z E R T Y rather than Q W E R T Y as in the most popular keyboard layout. Also has a range of accents and diacritics.
Backing storage	The media used to store information and programs needed to carry out a task using a computer-based system; usually of greater volume but required less urgently for immediate processing than *main memory* but is accessed for updating and retrieval more frequently than *archival storage*. *Magnetic media*

(Backing storage cont.)	are the most popular; *bulk* and *mass* storage/memory are frequently used as synonyms.
Bandwidth	Measurement of the telecommunications capacity of a transmission link. The 'wider' or 'higher' the bandwidth, the greater is the volume of information that can travel along it, either in the form of more channels or in carrying a particular type of information; TV pictures, for example, require a higher bandwidth than provided by traditional copper telephone lines.
Baud	Measurement of transmission capacity; equivalent to *bps*.
BCD	A code used in computer systems which uses six *bits* to represent all digits and alphabetic characters; abbreviation of Binary Coded Decimal.
Bi-directional printing	The printing of successive lines from left to right then from right to left, thus eliminating the time taken by conventional carriage-return motion.
Binary	A way of representing information and performing arithmetic calculations using a 'two state' representation (eg 0,1; on/off).
Bit	Contraction of *bi*nary digi*t*. A bit can have one of two binary values, usually represented as 0 and 1. See **Byte, Digital.**
Bits per second	A measure of the transmission rate applied to the communication of *digital* information; often abbreviated to bps.
Bootstrap	Procedure required to initiate a computer-based system (such as a word processor) in certain circumstances. See **IPL.**
bps	Abbreviation of bits per second.
Bubble memory	A form of compact, high-capacity mass memory; information is stored on tiny magnetic bubbles within a solid medium.
Bulk storage	See **Backing storage.**

Byte 8 *bits*; the basic internal unit used to represent information within most computer systems.

Calculator mode The ability available with some word processors to perform calculations on figures appearing in the text, such as column totalling. Also known as arithmetic mode.

Cartridge disk See **Disk**.

Cassette tape A *backing store* medium which can hold up to about 80,000 characters on magnetic tape; similar to the cassettes, originally developed by Philips and used for music.

Cat Device developed by Xerox Corp to enable the *cursor* to be controlled by the movement of the operator's finger on a touch-sensitive pad in front of the display screen.

Ceefax BBC version of *teletext*.

Centring The ability to use a word processor to automatically (or semi-automatically) centre headings, titles, blocks of text, tables and decimal points within a given space.

Character set The range of characters available for a display or printer, such as alphabetic characters, digits, punctuation marks and special characters (@, £, & and so on). See **Font, Pitch, Typeface**.

Chip See **Silicon chip**.

chps Abbreviation of *ch*aracters *p*er *s*econd; measurement of printer speed. Alternative abbreviation cps is sometimes used.

Closed User Group (CUG) A group of users of a *viewdata* system who have access to *frames* of information which are not available to the public or other users.

Close-up Basic display-based word processing facility. After a character, word or *string* of text is deleted, the resultant blank space is filled in by moving back the text appearing after the blank; the closed-up text is automatically placed in the required format.

Columnar working	Word processing capability allowing information to be moved about and manipulated as columns. See **Calculator mode.**
COM	Abbreviation of *computer output to microfilm*. Output from a computer is used directly or indirectly (by first being transferred to magnetic tape) to produce *microfilm* or *microfiche*.
cps	Alternative to chps as abbreviation for characters per second.
crash	Colloquialism indicating a breakdown in the system.
CRT	Abbreviation for *cathode ray tube*. See **VDU.**
CSA	Abbreviation for *Computing Services Association*.
Cursor	Pointer on a screen which identifies the position into which the next typed character will be placed. See **Ghost cursor.**
CWP	Abbreviation for *communicating word processors*.
Daisy wheel	A print mechanism consisting of a hub with spokes (called petals) at the end of which is a character, as at the end of manual typewriter print lever, available in a wide variety of typefaces and character sets which can be interchanged. Widely used to provide high-quality printed output with word processors.
Database	A computerised information store. Special database management *software* is available to help organise and provide simple access to information in the database. See **Information retrieval.**
Data entry	See **Input.**
Data processing	Computer-based information processing with a great deal of numerical manipulations. Contrasted with *text* processing which primarily involves word and text manipulations as in most office systems.

Dictionary	Word processing capability which allows reference dictionaries to be created to assist the typing process; for example, dictionaries containing hyphenation rules, spelling aids or common translations.
Digital	Storage, transmission and processing of information in the form of a digital code. *Binary* digital information techniques are used for electronic office systems.
Directory	Word processing facility. Contains information, such as document names and lengths, used to manage documents held on backing store.
Disaster quotient	Measure of the reliability of a system; how often the text in a word processing has been lost irretrievably and the quantities lost.
Disc	Alternative spelling for disk.
Disk	Backing store medium, which works on similar principal to music discs, ie information is stored on tracks and can be retrieved by placing an appropriate reading device in the appropriate track. See **Diskette, Floppy disk, Hard disks, Magnetic media, Rigid disks.**
Diskette	See **Floppy disk.**
Display	See **VDU.**
Document assembly	Creating a document by assembling a mixture of text already stored in a word processor and freshly keyed-in text. See **Standard text.**
Document merge	See **Document assembly.**
Dot matrix	See **Matrix printers.**
Dump	Regular dumping of information from *backing* store helps to provide a security safeguard in case information is corrupted or destroyed.
EBCDIC	One code which uses a *byte* to represent all digits and alphabetic characters; abbreviation of *e*xtended *b*inary *c*oded *d*ecimal *i*nterchange *c*ode. See **BCD, ASCII, BCD.**

Editing	The process of checking and revising text using a word processor. See **Text editor.**
Electronic filing	Information which conventionally has been held on paper in filing cabinets is stored in machine readable form in a way that users can easily look up information
Electronic mail	A communication between two or more people in which the transmission is, at least in part, via electronic equipment; eg, the telephone, telex, *facsimile*, communicating word processors and computer-based message systems.
Electronic mail box	See **Store and forward.**
Electronic office	A general term used to describe what offices will be like with the increased use of electronic equipment and the resultant changes in office procedures and organisational methods and structures.
Electrosensitive	One form of *non-impact* printer; specially treated conductive paper is sensitised to produce the printed image.
Enhanced typewriter	See **Automatic typewriter.**
Executive terminal	*Workstation* designed for use by managers and professional staff.
Facsimile	A method of transmitting images from one point to another; acts like a 'telecopier' with the image of a document created on a receiver at a different location to the place where the document is placed in the device. See **Electronic mail.**
Fax	Abbreviation of facsimile.
Fibreoptic cable	Cables made up of hair-thin glass fibres through which lightbeams can be sent to carry information (or conventional lighting); a high *bandwidth* transmission method.
Floppy disk	A thin, flat, circular piece of plastic coated in magnetic oxide and contained within a protective envelope, similar to a flexible 45rpm record. Used as a *backing* store medium with

(Floppy disk cont.)	word processors and other computer systems. Sometimes called a diskette. See **Hard disks.**
Font (or fount)	A particular style of *typeface* design and print size.
Form design	Word processor capability for designing and storing document formats. Specifies areas to be used for text and information to be input by the operator, possibly incorporating automatic tabbing and *stop codes.*
Frame	One screenful of information; particularly used with reference to a Prestel page (24 rows of 40 characters).
Full page display	A screen capable of displaying at least 66 lines at 80 characters per line; most screens for word processors display far fewer lines.
Ghost cursor	Used on some word processors when an ancillary function is initiated during data entry or text *editing.* The main *cursor* is reduced in intensity and a temporary ghost cursor is created to carry out an ancillary function.
Golf ball printhead	An easily changeable metal sphere carrying raised characters; used with electric and some *automatic typewriters.*
Graceful degradation	The ability for a system to carry on working, probably at reduced performance, despite component failures.
Hard disks	A rigid magnetic oxide-coated disk usually provided in a cartridge container; higher capacity (many millions of characters) than *floppy disks.* See **Backing storage, Disk.**
Hardware	The physical equipment in an office information system.
Holographic store	New development in *mass* storage where information is held on 'film' and is read using a laser; higher capacity than magnetic media.
Housekeeping routines	Word processing capabilities which allow specific documents to be removed, deleted or rearranged on the storage media; this enables several

(Housekeeping routines cont.)
documents to be handled within the same system.

Hyphenation
Operator or *program* controlled method used to decide where to insert a hyphen when a word is broken at the end of a line. A *dictionary* of hyphenation rules is available on some systems.

Impact printers
Printers which create an image by some form of physical contact on paper. See **Daisy wheel, Golf ball printhead, Matrix.**

Information provider
Used in relation to Prestel: an organisation that provides information to be stored on the Prestel database to be accessed by users.

Information retrieval
Automatic search and extraction of information from computer *databases*.

Ink jet
Non-impact printing technique which creates characters as a matrix of dots produced by the high speed 'squirting' from ink jets.

Input
Process of entering information into a computer system.

Integrated workstation
See **Multi-function workstation.**

Intelligence
See **Microprocessor.**

Intelligent copier
A computer-based system which combines the capabilities of a word processor, *fax* and *OCR* techniques.

IPL
Abbreviation of *i*nitial *p*rogram *l*oader. Required to initialise systems before loading the software that will carry out required functions. With some systems, needed each time the word processor is switched on.

Justification
Process of ensuring that there is an even right hand margin; automatic *hyphenation* is likely to be needed.

K
Strictly speaking, abbreviation for 1024 but sometimes used as an abbreviation for 1000.

Keyboard
Means of typing *input* to a computer-based system or preparing a typed document. See **AZERTY; Keypad, QWERTY.**

Keypad
Compact keyboard, typically containing only the digits (0 to 9) and a few control keys.

Light pen	Handheld device to *input* to a computer-based systems; can be used to read bar codes on products or to 'point' to locations on a display screen.
Local network	A means of interlinking electronic office devices within a building, usually using a *wideband* cable link. See **Ring network.**
M	Abbreviation of Mega (1 million)
Magnetic card	A backing store method popular with early automatic typewriters and word processors; usually consists of a mylar film card, just over 3″ × 7″, coated in a magnetic film and capable of holding about 5,000 characters equivalent to about a page of typescript.
Magnetic media	Most popular *backing storage* technique for holding encoded digital information in the magnetic surface of various media. See **Cassette, Disk, Magnetic card, Magnetic tape, Bubble memory** — all use magnetic techniques.
Magnetic tape	Medium for storing information; tape widths range from ¼″ to 1″ and tape feed mechanisms include closed cassettes and cartridges, open reel-to-reel and endless tape loops. Tape has to be read serially to reach a particular item, as trying to find a particular track on a music cassette.
Mainframe	A large general purpose computer usually housed in an air conditioned computer room and controlled by full time operators. Used to distinguish the relatively larger cost and size of such systems compared to *mini* and *micro* computers, although there is overlap between these categories.
Main memory	Synonym for main storage. Usually a high speed electronic device, such as a microelectronics chip. Used to hold the programs and information which are required to carry out the current task. See **RAM, ROM.**
Margination	Word processing capability for controlling the start and end positions of lines of text.

Mass storage	A high capacity storage device; sometimes used as a synonym for backing or bulk storage but some media such as magnetic cards do not fall into the mass memory range, which is usually measured in millions of *bits*.
Matrix printers	A means of forming characters on a screen display or in a printed form as a pattern of small dots. A 7×9 dot matrix grid is a typical matrix definition; sharper images can be created by having more dots.
Memory	Synonym for storage in computer-based systems; usually used in relation to *main* memory.
MICR	Abbreviation of *m*agnetic *i*nk *c*haracter *r*ecognition; a method of entering information into electronic processing systems, as with the numbers at the foot of a cheque.
Micro chip	A sliver of silicon, about $\frac{1}{4}''$ square, which can contain many integrated circuits that comprise a computer processor and/or memory. See **Microcomputer, Microprocessor, RAM, ROM.**
Microcomputer	A small, low-cost computer; used as a relative description compared to larger systems, *mainframes* and *minicomputers* although there is overlap between categories. A microcomputer on a chip is feasible but usually there is more than one chip per microcomputer. Must contain a processor, memory and input/output controller.
Microfiche	A 'card' form of *microfilm*, typically about $4'' \times 6''$ which can contain many frames of information. See **Archival storage, COM.**
Microfilm	A fine-grain, high-resolution film that contains a reduced image of a document which can be searched using a microfilm reader. See **Archival storage, COM.**

Microprocessor	The circuits which perform arithmetic and logical functions; essential to any computer-based *intelligent* device, such as a word processor, together with the necessary memory and input/output capabilities.
Microwriter	A portable handheld device for authors to record text using a five finger keyboard and a strip display.
Minicomputer	Generally smaller than what is now called a *mainframe* but larger than a *microcomputer*, although there is a great deal of overlap; often incorporated as part of a 'packaged' small business system or word processor.
Modem	A device which translates between digital information and the analogue wave transmission used on traditional telephone lines; one is needed at each end of a telephone line when communicating via analogue links between computer-based systems. Abbreviation of *mo*dulator *dem*odulator.
Mouse	A device which is used to move a *cursor* on a display and to initiate actions by moving the mouse unit physically in the direction corresponding to the movement of the cursor.
MTBF	Abbreviation of *m*ean *t*ime *b*etween *f*ailure; a measure of reliability of a system or device.
Multi-function workstation	A *workstation* from which a variety of electronic office functions can be carried out, such as text, data, voice and image processing. Also called integrated workstations.
Non-impact printing	Printing techniques which create an image in paper without involving physical contact between the printer and the paper. See **Electrosensitive, Ink jet** and **Thermal matrix.**
OCR	Abbreviation for *o*ptical *c*haracter *r*ecognition. Technique of inputting information directly from typed

(OCR cont.)	documents into computer code; many OCR devices can read only a limited range of *typefaces*. See **MICR.**
Office of the Future	A general term for the *Electronic Office*.
OOF	Abbreviation of *Office of the Future.*
Optical fibre	See **Fibreoptic cable.**
Oracle	Independent Broadcasting Authority's version of *teletext*.
Output	Presentation of results after computer processing; printers, screens, COM are examples of output devices.
Pad	A pressure sensitive device which interprets handwritten numbers and characters as they are written for direct *input* to a computer-based system.
Pagination	Word processing capability which automatically controls page usage and numbering of a multi-page document.
Paper tape	Tape punched with holes; early storage and input medium.
Password	See **Security.**
Personal computer	Low-cost computers designed primarily for personal home and educational use. Usually uses a domestic TV screen as the display. With suitable additional software and hardware capabilities, personal computers can be used for business applications, such as word processing. See **Microcomputer.**
Petal printer	See **Daisy wheel.**
Pitch	Number of characters that can be typed per inch.
Point	Unit of measurement specifying the height of a *typeface*.
Preprogrammed correspondence	Similar to *document assembly*.
Prestel	The trade name for British Telecom's *viewdata* service.
Processor	The part of a computer concerned with controlling the information processing needs of the user as instructed by *programs*. See **Microprocessor.**

Program	A detailed set of instructions to carry out specific information processing tasks. When stored in main memory, programs control the operations of the processor. See **Software.**
Programmed corres-pondence	Word processing capability which allows text to be stored for subsequent inclusion in a document; once inserted into the appropriate position in the document, the pre-stored text can be revised as required. See **Document assembly.**
Proportional spacing	The capability to vary the space taken up by a character, for example, the letter 'i' takes less space than 'm'.
Protocol	The well-defined routines and standards used when transmitting information from one electronic device to another.
QWERTY	Standard keyboard layout for English language applications; the letters Q W E R T Y are the first characters in the top alphabetic row.
RAM	Abbreviation for *r*andom *a*ccess *m*emory; a form of main memory which is used to store information that needs to be altered, such as text which is being input or revised.
Raster scan	A technique for generating characters on a screen via an electron beam focused on a phosphor screen as a matrix of illuminated dots; typical technique used in standard TV sets. See **VDU, Vector generation.**
Repeat key	When pressed simultaneously with a normal character key, causes that character to be repeated as long as both keys are depressed.
Restore	Recreating the computerised files as they were before an error or breakdown occurred. See **Dump.**
Reverse video	Used to highlight particular portions of a screen; reverses the normal display contrast — if a bright character is normally displayed on a dark background, reverse video will have a dark character on a bright background.

Rigid disks	See **Hard disks**.
Ring network	A form of *local network* based on the same idea as an electric ring main; electronic devices can plug into a 'ring' cable to interchange information.
ROM	Abbreviation for *read only memory*; a form of main memory used to hold information that does not need to be changed, such as programs that have been fully tested.
Screen	Part of a *VDU* or word processor which displays information as it is keyed in and presents output and responses from computers.
Scrolling	As a word processor or VDU screen can usually display only about half a page of text at once, scrolling is needed to scan other parts outside the screen *frame*. With scrolling, the VDU acts as a 'window' on the document, moving up and down (vertical scrolling) or from side to side (horizontal scrolling).
Search	Word processing capability which enables the location of a defined sequence of characters to be discovered and presented to the operator.
Security	Systematic techniques to prevent unauthorised access to information within an electronic information system. Could take the form of physical locks and/or protection by *password*, with the software allowing access only to someone who knows the required password.
Shared logic	See **Shared resource**.
Shared resource	Word processing systems in which more than one *workstation* can share processing, storage and printing resources.
Single-line display	See **Strip window**.
Software	Generic name for *programs*.
Split screen	The facility to divide a screen display so that it can show text from more than one document or page at the same time.

Stand alone	A self-sufficient word processor that does not need any external resources to fulfill its purpose; includes its own keyboard, screen, printer, main store, backing store and processing capabilities.
Standard paragraphs	See **Standard text.**
Standard text	Frequently used text can be stored for use in a variety of documents; each section of standard text has an identifier that is used to recall a copy of it for merging with other standard or newly input text. See **Document assembly, Preprogrammed** and **Programmed correspondence.**
Stop code	Used to position the cursor automatically when a section of text is displayed, say at the start of a name field in a form. The same term is used in printing as the code which identifies where printing must stop, for example to change paper or the print-head. See **Form design.**
Storage	Techniques and medium for holding information in a *digital* encoded form for subsequent retrieval and computer processing. See **Backing** and **Magnetic storage, Memory.**
Store and forward	An *electronic mail* service which enables a message sent from a *workstation* to be stored on computer media until the recipient at a *workstation* asks it to be forwarded to his or her workstation; avoids delays caused by 'engaged' or 'not in' responses.
String	A contiguous sequence of characters.
Strip window	A single line display; typically up to 24 characters, which is sufficient to display the last few words typed or to locate a point in the text for revising.
Tab rack	A message line at the top of a screen which defines tab and margin positions; equivalent to the metal tab rack at the back of a traditional manual typewriter.

Tablet   A special surface used as the base of an input device; the image data tablet, for example, recognises handwriting by monitoring the current through a pen used with the tablet.

Tape   See **Magnetic tape, Paper tape.**

Telecommunications The electronic transmission of information; either in an analogue form, as with sound wave transmission for telephone lines, or *digital*.

Teletex   'Super telex' offering higher speeds, better quality output and more editing and processing capabilities than traditional telex; international telex standards have been developed.

Teletext   Information dissemination system which broadcasts hundreds of pages of text and graphics, like a TV program; pages can be selected for display on a modified domestic TV set. See **Ceefax, Oracle, Videotex.**

Terminal   A device linked to a computer for entering and receiving information processed on a computer. See **VDU.**

Text editor   Software incorporated into a word processor or a software package on a general purpose computer to carry out *editing* functions.

Text processing   Computer-based processing, storage and transmission of primarily textual information; overlaps with, but distinct from, *data* processing. See **Word processing.**

Thermal matrix   Non-impact printing technique using a special paper which is sensitised by heat.

Tree structured   Method of organising a *database* similar to a family tree; information is retrieved by searching down each branch until the required item is found. Used in Prestel and other viewdata systems.

Typebasket   Print mechanism of old-style typewriter with raised character at the ends of levers.

Typeface   A particular style of character design. See **Character set, Font, Pitch, Point.**

Umbrella organisation	An information provider for Prestel who coordinates the needs of other organisations wishing to provide information.
VDT	Abbreviation for *v*isual *d*isplay *t*erminal. See **VDU**.
VDU	Abbreviation for *v*isual *d*isplay *u*nit. TV-like screen (usually using a *CRT*) which displays characters as they are typed on an associated keyboard and which presents *output* from computer processing. A screen-based word processor is an example of a VDU. See **Cursor, Raster scan, Tab rack** and **Vector generation.**
Vector generation	A technique for creating characters on a screen as a collection of lines rather than a dot matrix; of particular use in screens used for graphical displays. See **Raster scan, VDU.**
Videodisc or videodisks	*Mass* storage medium similar to videodiscs used for video entertainment; can enable many thousands of documents to be stored and retrieved. See **Electronic filing.**
Videotex	International generic term for information dissemination systems. Viewdata is an *interactive videotex* service and teletext is a *broadcast videotex* system.
Viewdata	Generic term for systems using relatively simple equipment (typically a modified domestic TV and keypad) to access computers containing thousands of information pages; users can leave messages on the system, such as placing an order or booking a seat. Developed by the British Post Office; *Prestel* was the first public viewdata system. See **Tree structure, Videotex.**
Voice recognition	Ability to recognise spoken words automatically and translate them into a suitable form for electronic storage and processing.
Voice response	Computer output in the form of a voice composed either of artificial synthesised sounds or by linking

(Voice response cont.)	together words and phrases recorded by people.
Wideband	Telecommunication channels that can carry a high volume of information. See **Bandwidth.**
Widow	Single line left on a new page or a single word on a new line.
Word processing	Application of computer technology to the typing process; can be carried out on purpose built *word processor* or a general purpose computer with suitable *software* and *hardware.* See **Editing, Electronic Office, Text processing.**
Word processor	Purpose built equipment to carry out word processing tasks. See **Editing, Keyboard, Screen, Shared resources, Stand alone.**
Workstation	A device or a desk unit containing many devices from which people carry out work using electronic information handling techniques. See **Executive terminal, Multi-function workstation, VDU, Word processor.**
Wrapround	Facility that allows text to be entered without worrying about the position of the end of line; the system automatically starts a word on a new line if insufficient space is available for it on the current line.

Appendix 2
CSA study—text processing strategy studies

Background to the studies

In 1978, the UK Department of Industry commissioned the Computing Services Association (CSA) to carry out a study called the Implications of Text Processing. The results of the study form the source information for this book.

The study was divided into three phases. Background research into the technology and its use was undertaken in Phase 1, which concluded with the production of a report and the development of a methodology for identifying and developing recommendations for an organisation to handle its text processing needs.

This methodology was applied in Phase 2 of the project, the Text Processing Strategies. During this phase, individual detailed studies were undertaken for ten major organisations in the UK.

The final part was to prepare a summary of the results and the implications for other users of text processing equipment and for equipment manufacturers and computing service companies. This book is one outcome of Phase 3.

Most of the initial research was carried out during 1979 and subsequently updated. The total cost of the study was about £300,000, which was shared between the Department of Industry and the organisations participating in it.

The organisations studied

Abbey Life Assurance
British Institute of
  Management
British Leyland
Citibank
Knight Frank and Rutley

Metal Box
Paymaster General's Office
Royal London Mutual
  Insurance Society
Spillers Ltd
Thames Water Authority

These organisations were chosen as representing a good cross section in terms of size, area of activity and existing computing and electronic office experience, although there was a bias towards organisations with a relatively high requirement for paper handling. Word processing experience ranged widely, from zero up to companies that have been in the forefront of the field and have used large numbers of word processors for many years. The most common level of experience encompassed one or two quite basic word processors, usually without a screen, and less than two years' experience with word processing.

Data processing experience also varied widely. Some organisations' only access to a computer was through a bureau while others had several large mainframes with private communications networks. In many respects, data and word processing have followed independent paths and the companies with the least data processing experience were not the same as those with the least word processing experience.

Activities covered in detail in individual organisations included head office functions, such as a personnel and finance, as well as local branch and site activities. Half the organisations employed less than 200 staff, while the other half employed over 1,000, with up to 50,000 in the largest.

It is recognised by the study team that, while a sample of ten is sufficient to give a general impression, it is too small to derive a very detailed picture. Wherever average statistics are quoted they are usually an average of the values for each company without applying any weighting to account for different company sizes and sample sizes between companies. In most cases, only parts of an organisation were examined in detail.

To preserve the confidentiality of company data, the

results were presented in a combined form, not as individual case studies. Each organisation, however, has been given its own detailed strategic report.

The CSA consultants who carried out the study

The study team was comprised of consultants from seven members of the CSA Text Processing Group: Arthur Andersen & Co, Communications Studies and Planning Ltd, Langton Information Systems Ltd, Logica Ltd, Pactel, Peat Marwick Mitchell & Co and the P-E Consultancy Group. Colin Leeson of Langton was the project leader, Diana Duggan of Logica (who subsequently moved to Urwick Nexos) was technical leader and Willie Jamieson of Arthur Andersen was marketing manager.

The CSA represents the major UK computing services companies. The following are brief profiles of the CSA members involved in the text processing study.

Arthur Andersen & Co

A partnership of British citizens which is able to draw on the research and experience of associated firms throughout the world. Has close links with associates in the US who have in-depth experience of text processing systems. The management consultancy division in London has over 100 professional staff experienced in the design and implementation of computer-based information systems. The firm has experience of working with other consulting firms and of running project teams made up of people from different organisations.

Communications Studies and Planning Ltd

An international professional services organisation specialising in communications and information technology. It offers consultancy and research services to suppliers and users of these technologies, including governments and telecommunications carriers, and contributes to the formulation of public telecommunications policy. CS&P's consultants undertake advisory and analytical work across the whole of its technological field for clients who range from public bodies to the newspaper industry and manufacturers of office equipment. It has carried out studies for British Telecom on electronic mail, viewdata systems and digital networks.

Langton Information Systems Ltd

The company has consciously set out to blend information science, text handling and data processing skills. It has pioneered a number of approaches in the integration of text and graphics and in the data processing techniques for handling both of these. It has built systems using these techniques for organisations including aerospace, earthmoving equipment manufacturers, government, volume car production and automative equipment, in several countries. It has also carried out market research studies, for example on the industrial uses of Prestel for British Telecom and the uses of information retrieval packages for the British Library.

Logica Ltd

The largest group in the UK involved in consultancy, implementation and training in the area of office text processing. The company also has considerable experience in the related areas of communications and information retrieval. Clients in the text processing area include Unilever Ltd, the British Steel Corporation, BL Cars, BP Trading, Shell UK, Lloyd's of London Press, the British Library, BBC, British Railways Board, Reuters, Norwegian Government, British Airports Authority, United Nations, Wellcome Foundation, NCR, Monotype Corporation and Linotype Paul.

Pactel

A subsidiary of PA International, the company has been working in the text processing field for many years. It offers a balanced range of specialist technical and business skills as an integral part of PA's comprehensive consultancy service. Specific areas of assistance provided include audits, strategic reviews, technical feasibility studies, implementation and market studies for suppliers. It has advised many UK and international companies on applications of text processing, including helping an insurance broking organisation to implement complex systems, and has also carried out high level research into electronic office systems.

Peat Marwick Mitchell & Co

One of the world's largest firms of management consultants and public accountants. The firm has a professional consulting staff comprising data processing specialists, economists, accountants, management scientists, engineers and marketing experts. Apart from academic and professional qualifications, most of the staff have had considerable practical experience at senior level in industry, commerce and government. It has carried out a variety of assignments in the UK and overseas. Clients have included government departments and nationalised industries as well as many of the largest industrial and commercial undertakings in the private sector.

The P-E Consulting Group

A leading independent firm of management consultants operating throughout the world. It serves industry, commerce, professional bodies, governments and international agencies. Professionally qualified staff includes accountants, architects, industrial relations, data processing and operational research specialists. It has been in business for over forty years, provides a very wide range of management consulting services, has worked in practically every industry listed in the Standard Industrial Classification, and every year operates in more than twenty different countries.

Results of the study

The study resulted in a number of specific recommendations covering the short, medium and long terms. These recommendations are summarised in Chapter 4 and emphasised throughout the remainder of the book. They were developed to match existing and foreseeable technological developments to real organisational requirements.

An analysis of the text processing activities within the ten organisations in the study assisted in ensuring that the recommendations remained close to real information needs. The remainder of this appendix describes the analysis of current office activities investigated in the study in more detail than was given in Chapter 4.

Typing working environment

Typing workloads and staff organisation were recorded in selected parts of each of the ten organisations, including both secretaries and typists. There was a high proportion of head office based activities.

The average typing rates recorded were three typing units (one and a half dense pages of A4) per typing hour. A typing unit is equivalent to 1584 characters or 22 lines of 72 characters. Wide variations (from 2 to 10 units) were found, depending on the skill of staff, the type of work and the equipment used. The two highest rates (6 and 10 units) included significant volumes of work produced on word processors. As shown in Table A2.1 below, staff spent on average just over half their time typing.

Table A2.1   Typing working environment

% of time spent typing	56
ratio of typists/secretaries to executives	1:6
% typists/secretaries serving one executive	19
% typists/secretaries working alone	53

Handwriting was by far the most common form of input to typing work followed by typewritten or printed text, audio and shorthand. In six organisations, over 60% of work originated in handwriting, rising to 90 to 100% for one organisation. Overall, just over half of the typing originated as handwriting with the remainder fairly equally shared between the other techniques, although there was more type-written/printed work than audio and more audio than shorthand. Companies with high proportions of audio and shorthand work were those with the highest proportion of staff working for a single boss and were evenly distributed in the productivity league.

Overall, about a third of the typed work was of a repetitive nature where word processors could be particularly helpful. Only one organisation had over 50% of its work falling into this repetitive category and three organisations had less than 20%.

Mail

Surveys were carried out on mail received by a central mail room or by a person whose job was to distribute mail. Mail which was passed directly by individuals to colleagues was not included. In all, about 40,000 items of mail were analysed.

On average, 3 items a day were received by each white collar employee, with two organisations averaging over 5 a day. In many cases, senior staff handled far more than the average value. Almost half the total mail received was from other employees of the same organisation, as shown in Table A2.2 over. Organisations employing more than 10,000 people had an average of 75% internal mail compared with 26% for smaller organisations.

Only one organisation used any electronic mail involving communicating word processors.

Less than 5% of all mail involved the use of telex and facsimile, the most basic form of electronic mail.

Although the high level of internal mail indicates that there is great scope within larger organisations for electronic mail, it was in the smaller organisations with a smaller proportion of internal mail that existing electronic mail activities were found.

Table A2.3 shows that the majority of mail was in typewritten form and a significant amount of handwritten material, but very few drawings. Table A2.4 illustrates the fate of mail received. The fact that more items were copied than thrown away indicates that paper files could be growing at the rate of about 3 items per day.

Almost half the mail was deemed to be essential to the recipient's work and only 5% was classified as useless. About 45% was sent to provide information rather than to enable the recipient to take action.

Table A2.2 Sources of mail (average %)

Source	%
Same company, same site	19
Same company, other site	29
External to company	52

Table A2.3 Form of mail (average%)

Form	%
Typewritten	61
Handwritten	30
Printed	8
Drawings	1

Table A2.4 What happens to the mail (average %)

Action	%
File	58
Pass on	34
Copy	8
Throw away	6

Table A2.5 Photo-copying (average %)

Form	%
Typed in the organisation	42
Printed matter	22
Other	35

Photocopying

A large proportion of photocopying was originally typed internally, as shown in Table A2.5 above.

Resource costs

Staff costs comprised over half office costs; accommodation was a higher percentage than equipment costs. On average, about 15% of office staff were typists or secretaries although in one organisation this rose to about 40% but was less than 5% in another. About half the office staff were professional and executive grades. Staff costs were more than 30% of turnover in two organisations and less than 10% in three others.

Index